Nexus Network Journal

LEONARDO DA VINCI: ARCHITECTURE AND MATHEMATICS

Sylvie Duvernoy, Guest Editor

VOLUME 10, NUMBER 1

Spring 2008

KIM WILLIAMS BOOKS

Nexus Network Journal
Vol. 10
No. 1
Pp. 1-206
ISSN 1590-5896

CONTENTS

Letter from the Guest Editor

5   SYLVIE DUVERNOY. An Introduction to Leonardo's Lattices

Leonardo da Vinci: Architecture and Mathematics

13   KIM WILLIAMS. Transcription and Translation of *Codex Atlanticus*, fol. 899 v

17   RINUS ROELOFS. Two- and Three-Dimensional Constructions Based on Leonardo Grids

27   BIAGIO DI CARLO. The Wooden Roofs of Leonardo and New Structural Research

39   SYLVIE DUVERNOY. Leonardo and Theoretical Mathematics

51   MARK REYNOLDS. The Octagon in Leonardo's Drawings

77   JOÃO PEDRO XAVIER. Leonardo's Representational Technique for Centrally-Planned Temples

101   VESNA PETRESIN ROBERT. Perception of Order and Ambiguity in Leonardo's Design Concepts

129   CHRISTOPHER GLASS. Leonardo's Successors

Geometer's Angle

149   RACHEL FLETCHER. Dynamic Root Rectangles Part Two: Root-Two Rectangles and Design Applications

Didactics

179   JANE BURRY and ANDREW MAHER. The Other Mathematical Bridge

Book Reviews

195   MICHAEL OSTWALD. *A Theory of General Ethics: Human Relationships, Nature and the Built Environment* by Warwick Fox

199   SARAH CLOUGH EDWARDS. *Inigo Jones and the Classical Tradition* by Christy Anderson

203   SYLVIE DUVERNOY. *Architecture and Mathematics in Ancient Egypt* by Corinna Rossi

An Introduction to Leonardo's Lattices

Among the architectural and mathematical treatises that flourished during the Renaissance period, Leonardo's codices deserve special attention. They are not didactic treatises, arranged in several books that must be read from the first page to the last, but information about the scientific research in the Renaissance flows from their pages, full of sketches and notes as from an endless font. The reader always bumps into something new or unexpected when going through the drawings, whichever codex or whatever page he is exploring.

The sketches that can be seen on folio 899 of the *Codex Atlanticus* illustrate the design of a roof system assembled from simple elements, and describe the building process of this system based on the weaving of wooden logs that will generate a vaulted roof covering a wide space without intermediate supports (fig. 1). These drawings are quite unique in the pages of Leonardo and no repetition has been noticed in other codices, but they are not unique in the vast amount of architectural literature of Middle Ages and Renaissance.

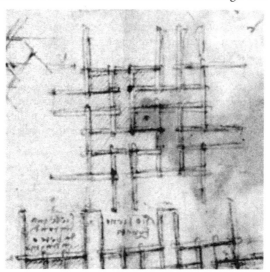

Fig. 1. Detail from Leonardo's *Codex Atlanticus*, folio 899

One of the drawings of Villard de Honnecourt's portfolio illustrates a building trick that can be considered to be an anticipation of Leonardo's research. The sketch by Villard shows a wooden floor built from beams that are all shorter than the dimensions of the room itself. The caption says: "in such a way you can work in a tower or a house with pieces of wood that are too short" (fig. 2). The beams are tied together according to a geometric pattern that makes it possible to cover of a span wider than the length of the beams themselves. We have no information whether Villard invented this trick together with some fellow builders, or if he inherited it from previous "know how", but the drawing is intended to transmit this technique to posterity and in fact it appears again, in drawing and in written description, in some treatises of the Renaissance.

Nexus Network Journal 10 (2008) 5-12   NEXUS NETWORK JOURNAL – VOL. 10, No. 1, 2008 **5**
1590-5896/08/010005-8 DOI 10.1007/ S00004-007-0051-0
© 2008 Kim Williams Books, Turin

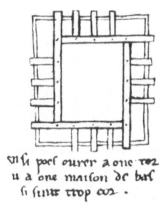

ꝟ ſi poeſ ourẜ a one ꝛoꝛ
u a one maiſon ꝺꝛ baſ
ſi ſunꞇ ꞇꞇop cꞷꝛ .

Fig. 2. How to build a floor with beams that are too short. From Villard de Honnecourt's portfolio (1225-1250)

In Book I of the *Seven Books on Architecture* by Sebastiano Serlio we can see a drawing that shows an application of the same technique: the system has been repeated and extended to build an even wider floor (fig. 3). Book I: *De Geometria* was first published in Paris in 1545, twenty-six years after Leonardo's death. Serlio's drawing looks like a combination of ancient problems and new solutions, but no explanation is given as far as design and science are concerned. Serlio only claims to present some possible solutions to inconveniences that often arise in the course of the architect's professional carreer. It is interesting to notice, however, that this practical trick of the trade appears in the more theoretical of the *Seven Books*, which deals with geometry.

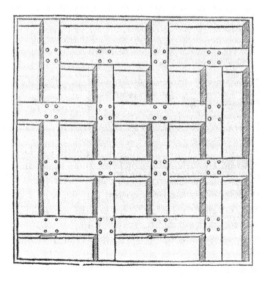

Fig. 3. How to build a floor of fifteen feet with beams that are one *braccio* short. From the *Seven Books on Architecture* by Sebastiano Serlio, Bk I: *De Geometria* (1545)

In the captions related to the drawings of folio 899 of the *Codex Atlanticus*, Leonardo does not claim to have discovered the construction technique that he illustrates. But while Villard and Serlio only deal with one single geometric pattern based on squares, Leonardo's investigates four different geometric patterns, not all of them based on orthogonality. In addition, since his beams are not abutted and nailed together they do not produce an horizontal surface, but rather a vaulted one. Because his notes are so specific in describing quantities and areas together with the details of the building process, it seems, however, quite certain that he actually took part in – or even directed – some building experiment of this kind.

The papers that are presented in this issue of the *Nexus Network Journal* are the result of a workshop that took place in Vinci, birthplace of Leonardo, in June 2003, sponsored by the Leonardo Museum and Library of Vinci and Kim Williams Books, dedicated to the study of "Leonardo's lattices".

The workshop was composed of two parts: first, theoretical studies, and then the experimental phase. For the first part of the project, the seminar, a team of experts in various fields and from various nationalities discussed the use of geometry in the architecture of Leonardo as found in his sketchbooks. Presentations were made by Biagio di Carlo, Sylvie Duvernoy, Christopher Glass, Vesna Petresin, Mark Reynolds, Rinus Roelofs and João Pedro Xavier. The papers in this present issue all grew out of the research for and exchange of ideas during the workshop.

The second part of the project involved the actual construction of domes based upon Leonardo's system. During this second phase, which acted as the verification process of the theoretical research, our group decided to build four vaults following the instructions and drawings of Leonardo. It was important for us to take the discussion from the realm of theory into the realm of practice, since the construction of the dome allowed the theory to be tested. The construction was directed by Dutch artist Rinus Roelofs, who has worked with Leonardo's system of bar grids since 1989.

Leonardo himself gives some starting instructions:

Sien legnami tondi, d'abete o castagni. Non sieno forati

(Let them be round logs, of fir or chessnut. Let them not be drilled with holes).

Our beams where four meters long, and in order to arrange them very regularly we first made four notches in each, to mark the precise position where they had to intersect with each other. The notches also helped prevent the beams from sliding, since they were not fixed to one another by either nails or ropes, but were only "woven". For our first experiment we chose the orthogonal pattern based on the composition of square and rectangular shapes (fig. 4). This pattern may adequately cover a square space if regularly expanded from the center in both directions, or it can be developed along a single direction to cover a rectangular space, or even form the structure for a kind of bridge.

Fig. 4. First dome completed

Unlike masonry vaults or cupolas, which are usually built from the exterior towards the interior, that is, starting from the supporting walls and proceeding towards the center of the space, this kind of woven wooden structure starts from the center and expands outwards, the vaulted form rising in proportion to the width or diameter of the covered space and the thickness of the individual elements: the thicker the beams are, the higher the dome will rise.

The second vault that we built was the one based on a geometrical pattern in which hexagons and equilateral triangles alternate. We later tried another kind of orthogonal pattern composed only of squares of two different sizes (figs. 5 and 6).

Fig. 5. Second dome completed

Fig. 6. Third dome completed

The rapidity of the building – and unbuilding – of the vaults allowed us to construct four of them in two days. The beams can be lifted and assembled by three or four men at most, no machinery being necessary, but the more the vault expands, the heavier it becomes and so the more difficult it is to lift and insert more beams at the edges to continue to enlarge the structure. The ultimate limit to the width of the vault is proportional to the strength and maximum resistance of the beams that touch the ground. Leonardo suggests doubling them in order to increase the stability of the structure:

> *ma con certezza si romperà li più deboli, li quali son li più carichi, e quelli che son piè carichi son quelli che toccan terra, che sostengano il tutto, li quali fien raddoppiati, che ne tocca a sostenere tre quarti di cantile*

> (but certainly the weakest ones will break, those which carry the greatest loads, and those carrying the greatest loads are those that touch the ground, which can be doubled, because it falls to them to carry three-quarters of the load).

Indeed, at one point one of our beams broke (one that was close to the edge but not actually resting on the ground). Part of the dome collapsed as a consequence, but not all of it, and we were able to repair the damage by removing the broken beam and inserting a new one, without having to dismantle it further.

Leonardo indicates that a vault covering a space of 45 *braccia* will have a height of 5 *braccia* (one Florentine *braccio* equals about 58.37 cm). The curvature of the vault is not very steep. The ratio of height to length is 1 to 9. The measurements we took of our own constructions confirmed this ratio. Some patterns produced a slightly lower curve than others, but overall the ratio of height to width varied between 1:8 and 1:10 (figs. 7 and 8).

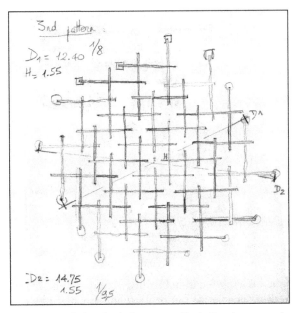

Fig. 7. Measurements of the third dome actually built: the space that has been covered can roughly been approximated to a square. The ratio between width and height of the dome varies from 1/8 to 1/9.5 according to the peripheral supports that are considered

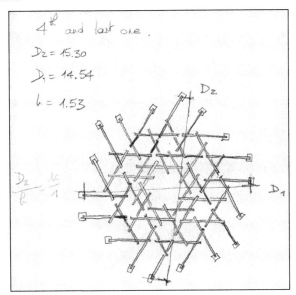

Fig. 8. Last but not least: geometric complexity increases in the pattern of the fourth built dome. Hexagons alternate with triangles and rhombuses. The dome is interrupted by a large central hexagonal oculus. The ratio between width and height – measured at the edge of the oculus – is close to 1/10

The fact that Leonardo speaks of beams that "touch the ground" shows that the experiment in which he participated was similar to ours. "His" vault too was erected on the ground and not on a peripheral masonry structure. But he also speaks of lifting the whole roof to raise it on some supports. For this operation machinery of ropes and levers appears to be required:

Debbon s'alzare tutti a un tratto colle lieve

(The whole should be lifted all at once with levers).

To complete the structure, this roof may be covered with fabric. Leonardo in his notes calculates the number of cloths necessary for a vault that covers a space of 45 *braccia*.

The whole building thus appears to be done with standard materials: small, identical timber beams, pieces of fabric *li quali ordinariamente si fan 30 braccia per ciascuno* (each of which is usually 30 *braccia*). The economy, ease and rapidity of this technology suggested to Carlo Pedretti, the major analyst of Leonardo's writings, that this kind of building was intended as some kind of temporary or emergency shelter. While this hypothesis is logical, nothing in Leonardo's own words of confirm the hypothesis, as he indicates nothing of the purpose and the function of such a structure.

We must consider this experiment as part of the general Renaissance research on relationships between mathematics and architecture. So far, those relationships have been investigated mainly with regards to the interaction between geometry and design, where geometrical shapes and patterns, together with their numerical proportions, guarantee the aesthetic result of the final architectural object. Here, in this experiment, geometry and mathematics are related to building technology, and Leonardo's concern obviously focuses on the role of geometrical shapes in structural stability.

Dome building is only one of the things that can be done with the Leonardo grids. The next step, by varying the basic building element, leads to building spheres, cylinders, and columns. Famous architects and engineers, such as Guastavino, Fuller and Snelson, would later study related structures.

We would like to take this opportunity to thank those who supported the 2003 Leonardo seminar and construction project: The Biblioteca and Museo Leonardiano of the city of Vinci, especially director Romano Nanni and librarian Monica Taddei, Laurent-Paul Robert for coordination of photography, and student helpers Lorenzo Matteoli and Alessio Mattu for help with the construction. And, of course, thank you very much to all those who participated for making the seminar and construction not only a learning experience, but fun as well.

Sylvie Duvernoy, Guest Editor

Kim Williams

Via Cavour, 8
10123 Turin ITALY
kwilliams@kimwilliamsbooks.com

Keywords: Leonardo da Vinci, Codex Atlanticus

Research

Transcription and Translation of Codex Atlanticus, fol. 899 v

Abstract. The basis for the 2003 seminar and construction project on Leonardo's roofing system was based on fol. 899v of the *Codex Atlanticus*. This paper is an transcription and translation to make that page more accessible.

Codex Atlanticus fol. 899v

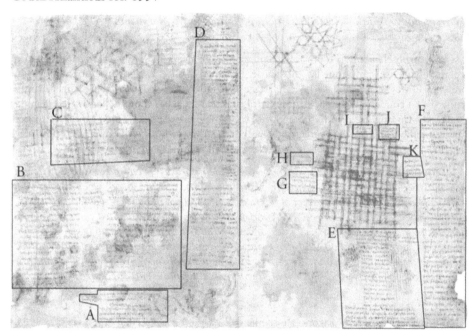

Left face

A.

La linea circunferenziale conterrà in sé tanto minor vano, quant'ella dista … di maggior lunghezza…spazio che s'allunga … la sua capacità come hanno

The circumference line will itself measure less by that amount that it lies from … of a greater length … space that lengthens …its capacity as have

B.

…bile e condensabile … da il mezzo dell'arco …strigner si condensa, l'altra … parte che da esso arco … fori si dilata e ra … ne porosità … ella resta di spezie di vacuo … il quale é potentissimo e al continuo … disfarsi, il che non po disfare … ancora l'aria condensata distende, e se tale arco sta mezzo carico, l'aria condensata caccia per le insensibile porosità la soperchia aria e tira per le opposite porosità altra aria che ristora il vacuo, e cosi l'arco resta sanza alcuna potenzia.

...ble and condensable ... from the middle of the arch ... tightening, it pulls, the other ... part of that arch ...remains outside dilates and ra ... porosity ... there remains a kind of vacuum ... which is most powerful and continuous ...come undone, which cannot come undone ... still the condensed air extends, and if such arch is half loaded, the condensed air pushes away, by means of the invisible porosity, the superfluous air and pulls, by means of the opposite porosity, the other air that restores the vacuum, and thus the arch stays without any force at all.

Ogni corpo elementato é poroso; adunque il legno é poroso e se sarà incurvato la metà...

Every body made of elements is porous; thus wood is porous and if it is curved by half...

Ogni corpo elementato é poroso e ogni porosità é piena in sé ... e l'aria condensata e rarefatta é violente. Adunque la porosità del legname, quando é incurvato, una parte se ne condensa e una se ne rarefa. La rarefatta e la condensata ne farà ritornare nella prima natura ... La condensata spigne e la rarefatta tira; seguita che 'l moto dell'arco si genera da soperchio a carestia, o voi dire da rarefazione e condensazione, e se l'arco sarà lasciato per alquanto tempo incurvato, la rarefazion si condenserà (attraendo per le insensibili porosità) e la condensazione si rarefarà, e cosi l'arco resterà colla acquisitata curvità.

Every body made of elements is porous and each porosity is complete in itself... and the condensed and rarefied air is violent. Thus the porosity of the wood, when it is curved, one part is condensed and the other is rarefied. The rarefied and the condensed [parts] will return to their original nature ... the condensed pushes and the rarefied pulls; it follows that the motion of the arch generates an excess and a deficiency, that is to say, from rarefaction and condensation, and if the arch is left for some amount of time in a curved shape, the rarefaction will condense (attracting by means of the invisible porosity) and the condensation will rarefy, and thus the arch will remain with the curve acquired.

C.

Moltiplica i lati delli 6 esagoni per ordine delli esagoni *a b c d e f*, e di': '6 vie 6 fa 36,' e tante son le travi di 2 braccia che vanno in tal componimento e non si conta l'esagono di mezzo, perché i sua lati son fatti delle dette 36 travi, e sarà il suo circuito 50 braccia di diamitro.

Multiply the sides of the 6 hexagons by the order of the hexagons a b c d e f, that is, '6 by 6 makes 36,' and that is how many are the 2-braccia long beams in this arrangement, and the hexagon in the middle is not counted, because its sides are made of the 36 beams already counted, and its perimeter will be 50 braccia in diameter.

D.

Questi 18 quadrati a moltiplicarli per 4, perché ogni quadrato é composto in sé di 4 travi, tu arai 72 trave, ma in vero non sono se non 45 travi, che restan 27 men che non mostra tal moltiplicazione. E questo nasce perché li 2 quadrati di fori *a* e *c* e li 2 quadrati di sopra [c]ontengan 4 travi, tutti li altri son di 3 travi; e li quadrati di mezzo, cioè a b dal quadrato di sopra in fori che ha due travi, tutti li altri quadrati son fatti da una trave sola.

These 18 squares are multiplied by 4, because each square is itself composed of 4 beams, you will have 72 beams, but in fact there are only 45 beams, that is, 27 fewer than the product of the multiplication. This is because the 2 outside squares a and c and the two upper squares contain 4 beams, all the others are of 3 beams; and the middle square, that is a b of the upper outside square has 2 beams, all the other squares are made with one beam only.

Addunque di 72 travi che ti dava la predetta moltiplicazione, ne son diminuite 21, perché la somma di tale trave non son se non 51.

Thus from the 72 beams that are given by the multiplication described are taken away 21, because the sum of the beams is precisely 51.

Di questa incatenatura quadrata a b c d f si debbe di ogni 5, 4 multiplicalli per 4 e della somma levare 4 e 'l rimanente é il vero numero delle travi che vanno in tale collegazione; come dire delli quadrati a b c d f, che son quadrati, tu li moltiplichi per 4 per sé … ogni quadrato fa 40 di 4 … e dirai: 4 volte fa 20; levane 4, resta 16; adunque 16 sono li travi che compongano …

For this square configuration a b c d f for every 5, 4 multiply by 4 and from the product take away 4, and the remainder is the true number of beams that go in that configuration; that is, given squares a b c d f, which are squares, you multiply each by 4 … each square makes 40 of 4 … and I say: 4 times [five] makes 20; taking away 4, leaves 16; thus 16 are the beams that make it.

Right face

E.

164 corde v'ha a legare a copresi in mentre che si fa, (a un tempo medesimo). Levane uno, e prova a levarlo in diversi lochi, e vedi quanti ne ruina e che varietà v'ha il numero de' ruinati, ruinando in diversi lochi; ma fa che sien bene legati a ciò che, rompendosene alcuni, non abbino a discendere a terra.

164 cords are to be tied and covered as it is made (at the same time). Take away one, and try taking it away in several places, and see how many fail, and how many different failures there are, failing in different places; but let them be well tied so that, when some break, they don't fall to the ground.

Ma con certezza si romperà li più deboli, li quali son li più carichi, e quelli che son piè carichi son quelli che toccan terra, che sostengano il tutto; li quali fien raddoppiati, che ne tocca a sostenere tre quarti di cantile.

But it is certain that the weakest will break, those that are the most loaded, and those that are most loaded are those that touch the ground, that support the whole; let them be doubled, since they have to support three quarters of the beams.

F.

Dove da' lati non si può attaccare corde.

On the sides you can't attach cords.

Questo é da coprire uno spazio di 45 braccia, dove non si volessi puntelli in mezzo.

This will cover a space of 45 braccia, where you don't want columns in the middle.

Questi travelli ovver cantili sono in tutto 84, de' quali 24 sostengano li 60, che ogni cantile ne sostiene 1 e ½.

These little beams, or cantili, *are 84 in all, of which 24 support the other 60, each of these supporting 1 and ½.*

Li cantili lunghi 10 braccia, che 4 cantili per filo, che fan 4(0) braccia; e poi v'é 3 spazi infra li intervalli delle lor fronti di 3 braccia e uno per ispazio, che fa 10 braccia. Adunque 50 braccia fa il tutto, che per l'arco che fa la volta del tutto, diminuisce 5 braccia, che resta 45 braccia di spazio coperto da tale copertura, sopra la quale si tira panni lani interrsegati, come si da … stan le intersegazion delli spazi, che son 22 panni, li quali ordinariamente si fan 30 braccia per ciascuno.

The cantili *are 10 braccia long, and there are 4 cantili per row, making 4(0) braccia; and then there are 3 spaces in the intervals between them that are 3 braccia and one per space, making 10 braccia. Thus 50 braccia is the total length, which due to arching height of the overall vault, is diminished by 5 braccia, so leaving 45 braccia of space covered by this covering, over which are pulled overlapping woollen cloths, as are used …are the overlaps of the spaces, which are 22 cloths, which are ordinarily 30 braccia each.*

G.

Ciascun di questi legni ha 4 *busi*, eccetto li 24 che posano in terra.

Each of these logs has 4 busi *(intersections?), except for the 24 that rest on the ground.*

H.

12 panni copre il tutto.

12 cloths will cover the whole.

I.

Non sieno forati.

Let them not be drilled (with holes).

J.

Sien legnami tondi, d'abete o castagni.

Let them be round logs, of fir or chestnut.

K.

Debbon s'alzare tutti a un tratto colle lieve.

It has to be lifted all at once with levers.

About the author

Kim Williams is the editor-in-chief of the *Nexus Network Journal.*

Rinus Roelofs

Lansinkweg 28
7553AL Hengelo
THE NETHERLANDS
rinus@rinusroelofs.nl

Keywords: Leonardo da
Vinci, grids, structural
patterns, tilings

Research

Two- and Three-Dimensional Constructions Based on Leonardo Grids

Abstract. In 1989 I made a drawing of a net on a cube, consisting of 12 lines/elements. They were connected in a way that, a couple of months later, I recognised them in 899v in Leonardo's *Codex Atlanticus*. I don't know which moment impressed me the most: my own discovery of a very simple and powerful connecting system or the discovery of the Leonardo drawings, which implied that my own discovery was in fact a rediscovery. What we see in Leonardo's drawings are some examples of roof constructions built with a lot of straight elements. These drawings can be 'translated' into the following definition: On each element we define four points at some distance of each other – two points somewhere in the middle and two points closer to the ends. To make constructions with these elements we need only connect a middle point of one element to an end point of another one in a regular over-under pattern. Out of the simple definition of the elements, I designed many different patterns for my so-called "+ - - +" structures: domes, spheres, cylinders and other models were made.

Introduction

In 1989, I began constructing domes using notched bars assembled according to a simple rule. This led me to explore planar constructions based on this rule using fixed length "notched" linear segments. I was able to create a wide variety of patterns. From certain of these patterns I was able to construct spheres and cylinders with notched curved rods without the use of glue, rope, nails, or screws.

On folio 899v of the *Codex Atlanticus* we find, among others, three patterns with exactly the properties of these bar grids. In view of the way in which the patterns are drawn, oblong forms that seem to lie on each other, the most direct interpretation is that here we are dealing with stacking constructions, built from straight rods. Making a model leads to exactly the domes that I experimented myself. So the conjecture that Leonardo da Vinci is the first inventor of these constructions seems justified, although we cannot be sure about this.

Thus, the name I gave to my bar grid construction system is "Leonardo grids", with which I was able to construct all kinds of structures out of simple elements using one single constructing rule. Most of the constructions I made where planar and static. However, the "Leonardo grid system" also makes possible the construction of non-planar and dynamic structures.

The system

The construction admits a simple description. We start with a number of rods, on each one determining four points as indicated in fig. 1. We call these points connecting points.

Fig. 1

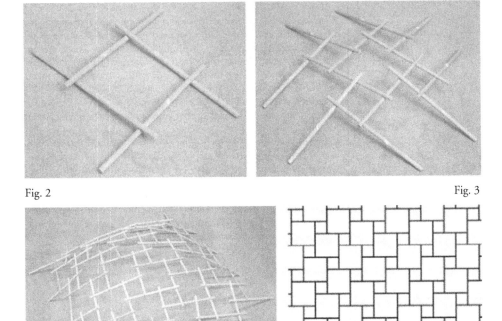

Fig. 2

Fig. 3

Fig. 4

Fig. 5

We distinguish two types of connecting points: end points (closest to the ends of the rods) and interior points (the remaining points). Each rod has two end points and two interior points. In constructing the dome we now apply the following rules: one of the endpoints of a rod is placed on a free interior point of a different rod. At the end all connecting points of the rods have to be used as a connection between two rods, except near the border of the construction.

Now the actual construction of the dome turns out to be a simple task. Beginning with four rods, as in fig. 2, we extend the construction by continually adding rods at the bottom (fig. 3). Since we add one rod at the time, on the outer edge, the dome can be constructed by a single person. The four rods with which we have started will rise automatically during the building process and, at the end, the dome, consisting of 64 rods, will stand on the ground, resting on only 16 rods (fig. 4).

Following the above construction process various patterns can be formed, each leading to a dome-like construction. In the sequel we will call the patterns that can be formed with the above rules "bar grids". The bar grid of the dome of fig. 4 can be drawn simplified as in fig. 5. In this form the drawing looks like a tiling pattern. However, we are not interested in the tiles but in the joints between the tiles. So we have a grid consisting of straight lines representing the rods. A first investigation into the various possible bar grids soon resulted into dozens of patterns, some of which are shown in fig. 6.

Fig. 6

From 2D to 3D

In the domes it is gravity that keeps the loose rods together. It follows that continuing the construction as far as a complete sphere is not possible. Yet it turns out that, using the above construction system, objects can be formed in which the elements themselves, instead of gravity, keep the construction together. For example, we can assemble a sphere from a number of rods, or more generally elements, without using connecting materials like wire or glue. The number of connecting points per elements and the connecting rules do not change. It is only the form of the elements that changes. For a sphere we use curved rods instead of straight ones.

A simple way to come to a design for such a sphere shaped construction is the following: in the bar grid of fig. 7 the midpoints of the hexagons are connected such that a pattern of triangles results (fig. 8). Eight of these triangles can be used to form an octahedron (fig. 9).

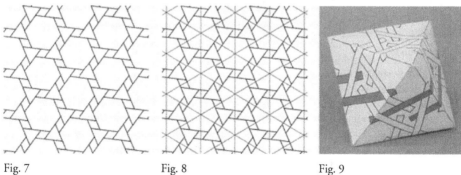

Fig. 7   Fig. 8   Fig. 9

On this octahedron we now see a grid consisting of 24 bars and this can be used as a design for the sphere of fig. 10. The form of the elements has been determined such that no tension arises in the sphere. Only when closing the sphere some elasticity from the elements is required. The relative position of the elements causes the sphere to stay in one piece: each of the elements is prevented from falling by other elements. For the sphere of fig. 11, which consists of 90 elements, the icosahedron has been used as an intermediate step so that pentagons occur in the construction.

Beside domes and spheres other shapes have been realized, such as cylinders and ovoids (figs. 12, 13).

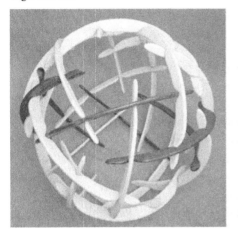

Fig. 10. Sphere, 24 elements sphere

Fig. 11. Sphere, 90 elements

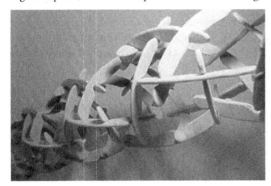

Fig. 12. Cylinder

Fig. 13. Ovoid

A real new step was made when designing some objects in which the inner space of the sphere is used too, as in fig. 17. This object has the form of two linked concentric spheres. The whole is a stable construction consisting of 24 elements. Each element is halfway (that is to say with two out of four connecting points) in the outer sphere and halfway in the inner sphere. The design was made by starting with two layers of Leonardo grids. The layers were placed above each other in such a way that after cutting each element in two parts, half elements from the upper layer could be connected to half elements of the under layer. The resulting structure again has all the properties needed for a Leonardo grid: each element is connected to 4 other elements in the right way: all endpoints connected to midpoints and all midpoints connected to endpoints. Half of the connecting points are on the surface of the inner sphere, the other half on the surface of the outer sphere. This is the first non-planar or real 3D result. And surprisingly enough, the total structure appeared to be stable.

Another approach was to start with interwoven patterns (fig. 14). A simple way to transform a flat pattern into a spherical construction is the use of polyhedron. When the pattern is hexagonal, the net of a icosahedron can be used. In the special case of the three interwoven patterns of fig. 15, the cut-off elements of pattern A will be connected to the cut-off elements of pattern B when folding the net into an icosahedron. And so with the cut-off elements of B and C and of C and A.

The result is real 3D Leonardo grid construction. And now all the connecting points are laying in the same spherical surface. In the pictures (figs. 16, 17, 18) you can see some variations.

Although this is a good approach it is hard to find a good set of interwoven patterns that can be used for this method.

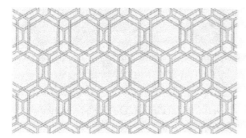

Fig. 14. Three interwoven patterns   Fig. 15. Icosahedron, plan

Figs. 16-18. Interwoven spheres

Infinite double layer structures

To go one step further towards the Leonardo grid space frames, I first tried to find a way to construct infinite double layer structures. Space frames can be built by connecting polyhedra in systematic way. With cubes you can fill the space. And when you look at the graph that represents the cube you will notice that all vertices have degree 3, which was a condition for the Leonardo grids. A cubic frame can be made as a Leonardo grid construction in three different ways (figures 19, 20 and 21).

A way to make a double layer structure is to connect these kinds of cubes as in fig. 22. This will result in non-planar infinite construction that has also dynamic properties. The elements can slide between certain boundaries and the total construction can be pressed together or stretched (figs. 23, 24).

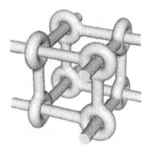

Fig. 19. Cubic frame A   Fig. 20. Cubic frame B   Fig. 21. Cubic frame C

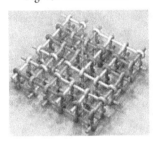

Fig. 22. Double layer   Fig. 23. Double layer   Fig. 24. Double layer structure,
structure   structure, pressed together   stretched

Rings and strings

In the double layer structures the basic elements, the rods with the 4 connecting points, are linked together to form bigger units. We can distinguish two kinds of these bigger units: rings, a closed concatenation of a finite number of basic elements, and strings a (open) concatenation of an infinite number of basic elements. Some examples of both categories are shown in figs. 25, 26 and 27 (rings) and figs. 28, 29 and 30 (strings).

The constructions needn't limited to two layers, as can be seen in figs. 31, 32 and 33. Here an infinite 3D construction is build with one type of string, which is a concatenation of basic Leonardo grid elements.

The question still is whether it is possible to make space frame constructions out of basic elements, which are not linked.

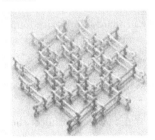

Fig. 25. Double layer structure   Fig. 26. Pressed together   Fig. 27. Stretched

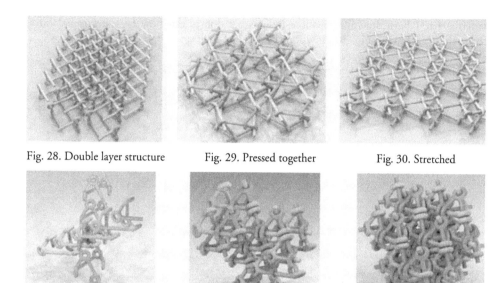

Fig. 28. Double layer structure   Fig. 29. Pressed together   Fig. 30. Stretched

Fig. 31. 3D Strings   Fig. 32. 3D String structure   Fig. 33. 3D String structure

Grid transformation

Another, and maybe even better way to construct real 3D structures based on Leonardo grids appeared to be the use of transformation of the basic Leonardo grid from 2D to 3D (figs. 34, 35). The process can be described as follows: we can start with any pattern in which we have a hexagonal hole. We now keep the 6 sticks around this hole connected and change the hexagon from flat to skew. This change will cause a transformation of the sticks, which are connected to the first 6 sticks. The six parallelogram shaped holes in the pattern will also be parallelogram-shaped at the end of the process. But one of the connections around the triangular holes will get loose. The resulting structure now can be used as a layer with which we can create space frames.

Fig. 34. Leonardo grid 2D   Fig. 35. Leonardo grid 3D

The discovery of this process lead to many designs of Leonardo grid space frames because the process could be applied on all the flat basic patterns. We can also start with the pattern with a square hole. In the same way the flat hexagon is transformed into a skew hexagon, a flat square can be transformed into a skew square. The over a hundred different patterns that I have drawn as possible designs for domes can now be transformed into Leonardo space frames. Fig. 36 shows one example of such space frames.

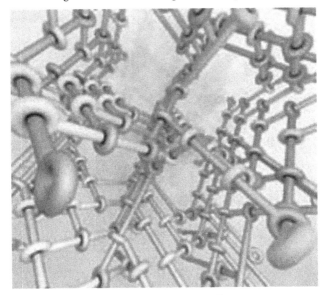

Fig. 36. Space frame grid

Dynamic space frames

Like the double layer structures, the Leonardo grid space frames also have dynamic properties. The sticks can be slid along each other and so the total construction can be pressed together or stretched. To show this we will go back to the skew hexagon first. In figs. 37, 38 and 39 you see three stages of the sliding process. And we can extend that to a complete layer, as in figs. 40, 41 and 42. In the total movement it looks as if there is some twist in the structure. The shrinking and growing of the structure is in a way a spiral movement. This can be best viewed in animation.

Figs. 37-39. Sliding: first, second, and third phases

Figs. 40-42. Sliding: first, second, and third phases

Dynamics in 2D

While studying dynamics in the 3D Leonardo grid constructions I also discovered an interesting new way of translating 2D Leonardo grids into dynamic structures. Looking again at the basic grids, you have to realize that there are two possible interpretations: you can either look at it as a construction built out of rods, or as a tiling, a pattern built with a set of tiles. The dividing lines then form the Leonardo grid.

Figs. 43-45. Transforming: first, second, and third phases

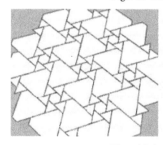 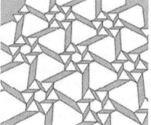 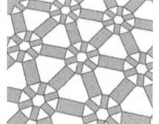

Figs. 46-48. Transforming: first, second, and third phases

One line in this grid represents the edges of 4 tiles: 2 big and 2 small tiles. So this one gridline can be seen as a set of 4 edges. Because the edges are alternately long-short-long-short, the set of edges can be seen as a parallelogram. In the original grid, the tiles are close to each other so the joints have area zero. But what will happen when we 'open' the parallelogram? The set of tiles then turns out to be a dynamic hinged construction. The Leonardo gridlines transform from lines to parallelograms to rectangles, and via parallelograms, to Leonardo gridlines again. And as a result of this process, a left-hand oriented Leonardo grid has been transformed into a right-hand oriented grid.

Because this process can be applied on any Leonardo grid, this leads to an enormous collection of hinged tile constructions. Most special is that this process also works on Leonardo grids with distortions, that is, Leonardo grids in which more then one pattern is used, and on Leonardo grids in which more then one length of grid lines is used. Some examples can be seen in the figures.

Of course, this can be better viewed in animation.

About the author

Rinus Roelofs of Hengelo, Netherlands, is an artist. He was born in 1954. After studying Applied Mathematics at the Technical University of Enschede, he took a degree from the Enschede Art Academy with a specialization in sculpture. His commissions come largely from municipalities, institutions and companies in the Netherlands, but his work has been exhibited further afield, including in Rome as part of the Escher Centennial celebrations in 1998. More information about his activities is available on his website, http://www.rinusroelofs.nl.

Biagio Di Carlo

Via Berlino 2
Villa Raspa, Spoltore
65010 Pescara ITALY
mail@biagiodicarlo.com

Keywords: Leonardo da
Vinci, reciprocal frames,
deresonated tensegrity,
geodesic domes,
Buckminster Fuller,
tensegrity, polyhedra

Research

The Wooden Roofs of Leonardo and New Structural Research

Abstract. The two types of spatial patterns reproduced in the *Codex Altanticus* fol. 899v can be deciphered in light of recent studies on reciprocal and tensegrity frames. For the construction of his wooden component roofs, Leonardo utilized two main modules: a grid of square modules and a grid of a tri/hexagonal module. Leonardo's drawings offer an opportunity to attempt a synthesis between the two structural systems, demonstrating the affinity that exists between the reciprocal frames used by Leonardo and the rigid tensegrities developed by Fuller. The continual observation, study and construction of models have permitted the verification of this hypothesis.

The framework idea reappears in a most ingenious system of dismountable system of "geodesic" roofing for vast areas of land, which can be seen in a later sheet of the *Codex Atlanticus*, f. 328 v-a of circa 1508-10, anticipating the daring constructions of Buckminster Fuller!

Carlo Pedretti [1978: 151]

Reciprocal frames

Reciprocal frames are three-dimensional structures, the beginning module of which must contain at least three sticks (the triangle is the first manifestation of a minimal surface) arranged in such a way as to form a closed circuit of mutually supporting elements. These structures permit the realization of any form whatsoever, so obtaining final configurations that are surprisingly stable.

Reciprocal frames are capable of supporting considerable loads. The eventual breaking of only one element jeopardizes the whole system, as is generally the case with synergetic structures. They can be rapidly constructed with local materials so that the result is particularly appropriate in emergency situations.

Reciprocal frames can be observed in the iris of the eye, in the aperture of a photographic camera, in the nests of some birds, in the art of basket weaving (especially Chinese and Thai), and in other varied examples.

Medieval Chinese and Japanese architecture

The reciprocal frames in wood used since the twelfth century in Chinese and Japanese architecture were destroyed by fire or allowed to deteriorate naturally with time, so that there remains little or no trace of the ancient traditions of construction. In more recent times we find them in the architecture of Ishi (the Spinning House), Kan and Kijima (Kijima Stonemason Museum), used particularly as roofing.

The Museum of Seiwa Bunrakukan by Kazuhiro Ishi (1992) was the object of a 1999 study on the part of groups of students of the University of Hong Kong, who had already had experience with reciprocal studies under the guidance of Ed Allen. As Gotz Gutdeutsch reports in his book *Building in Wood* (1996), Ishi said that he was inspired by the game of *waribashi* in which the sticks (chopsticks, usually used for eating) were joined by rubber bands in order to form houses, bridges and other objects. Another source of inspiration was the temples of Buddhist Chogen monks (1121-1206).

The puppet theatre has been part of Japanese culture since the sixteenth century. Each puppet was manipulated by three puppeteers in connection with the musicians and narrators. Even if the new types of media have largely supplanted this antique art, at Seiwa, on the island of Kyushu, it has been fortunately maintained. For the restoration of the theatre, the architect relied heavily on wood, in support of the regional industry for this material. The entire operation was notably successful.

The complex is divided into three principle areas: the auditorium, with a square plan and pyramidal roof; the theatre, with a rectangular plan; the exhibit hall, similar to a pagoda, connected to the auditorium by a covered walkway. In this room, metal joints were used to connect the large and elegant reciprocal roof.

One other architect deserving mention with regard to reciprocal structures is Tadashi Kawamata [Gould 1993]. The use of bamboo in the construction traditions of China should also not be neglected.

Rainbow Bridge. Another example of a reciprocal frame can be observed in the painting *Going up the river on the Rainbow* Bridge by Zhang Zeduan (twelfth century). In his book *Zhongguo Quiaoliang Shihua* (Taipei, 1987), Mao Yisheng reproduces a section of the Rainbow Bridge, demonstrating that the beams are placed so that they can be fixed in a reciprocal way. Realized for the first time in the province of Shandong, the Rainbow Bridge was one of the most important inventions during the time of the Chinese Song dynasty (960-1280 A.D.). It has the same place in Chinese culture as the Coliseum of Rome has in our own. According to Robin Yates of McGill University, the first foreigner to document that period of vigorous economic prosperity was Marco Polo, who arrived in China in 1275, at the end of the Song Dynasty. Before this dynasty, there were no printed books, gunpowder, compasses for navigation, paper money, restaurants, or bridges.

After many years of study and research, the engineer and historian Tang Huan Cheng recently reconstructed the Rainbow Bridge at full scale, collaborating with two experts from MIT, Marcus Brandt for the geometry, and Bashar Altabba for the structural engineering. The construction took place by starting on the two opposite sides of the river and connecting the whole structure with many timbers arranged in a reciprocal manner. Leonardo da Vinci has written that an arch is made of two weak halves that become strong when united. He also appreciated the way for spanning distances with short elements as illustrated in his three-dimensional grillage structures and temporary timber bridges. The principle used by Leonardo is identical to that used by the Chinese several hundred years before.

Other examples of reciprocal frames can be found in the wooden roofs designed by Sebastiano Serlio in the sixteenth century, and in the works of Villard de Honnecourt.

Hybrid systems

For Tony Robbin, "the use of hybrid systems constitutes a new constructive paradigm of resistance similar to that which exists in living organisms. The overloading of a structure is not only a waste of material but can be very dangerous." The Indian teepee is a hybrid

---

system based on the interaction of various independent systems that come into play only when they are necessary. That is, an active system exists for normal loads while the other systems become active as they are needed. The sticks of a teepee are a distant relative of a reciprocal structure. As happens in all tents, the covering fabric provides additional strength to the structure. In the summer the lower folds are opened to favor the air exchange and circulation.

Further studies on reciprocal frames have been conducted for the past fifteen years by John Chilton, Olga Popovic, Wanda Lewis and others. The results of their research have been published periodically in the magazine *International Journal of Space Structures*. For example, IJSS no. 2-3 (2002) contains the results obtained by students in the structural engineering course held by John Chilton and Wanda Lewis at the University of Warwick, who examined some models of reciprocal frames with a Testometric machine with a 100 Kn capacity, loading the models with weights. Chilton, Saidani and Rizzuto give the following definition of a reciprocal frame: "A Reciprocal Frame system is a three-dimensional grillage structure constructed of a closed circuit of mutually supporting beams. A number of RFs connected to each other at the outer end of each radiating beam results in the formation of a Multi-Reciprocal Element space structure". According to Chilton, one of the first to experiment with these kinds of structures was Emilio Perez Pinero.

Lincoln Cathedral (James Essex, 1762). Clearly described by John Chilton and Thibault Devulder of the University of Nottingham, the Gothic cathedral of Lincoln is one of the largest English cathedrals with a decagonal plan, constructed between 1220 and 1235 by master craftsmen and able carpenters. It is composed of two principle bodies: a structural base in dense stone of a diameter of some 21 m, and a height of some 13 m; and a wooden roof organized in two separate structures where, in the lower part, the heavy reciprocal structure realized by architect James Essex in the 1762 restoration can be noted. Essex broke with the current structural tradition by utilizing pine beams rather than oak.

Rigid Tensegrity Structures

I introduced non-resonant tensegrity structures – called "deresonated tensegrities" by Amy Edmondson (1986), and "Rigid Tensegrities" by Hugh Kenner (1966) – in the article "Le strutture tensegrali" published in *L'architettura naturale* 10 (2001):

> "Deresonated tensegrity domes" were Fuller's major interest in the last years of his life. Increasing the frequency of subdivision of the polyhedron of departure decreases the distances between the struts and gives importance to the thickness of the struts themselves, which can be dimensioned in such as way so as to permit adjacent struts to touch. This thickness can be calculated by taking into consideration the value of the respective geodesic arch. In this way the structure is without resonance (the tensegrity is deresonated), since the struts are no longer hung but touch each other and can be bolted together at their tangency points.
>
> The tension force otherwise visible in the preceding models becomes invisible in this type of structure. There is an evident reduction of the materials, that is, the number of struts of different lengths is reduced. In fact, there are only two different struts for a non-resonant 4v, in contrast to the eight struts necessary to construct the equivalent geodesic (p. 62).

In deresonated tensegrity structures the dynamic quality that permits the structures to oscillate from their position of initial equilibrium is blocked (or rendered non-resonant). Increasing the frequency of subdivision, the central corners near the struts are modified and tend towards a form that is less acute and nearer to spherical. The struts try to touch each other and can then be fixed with nuts and bolts which take the place of the tension cables. The resulting structure will be more robust and subject to very few bending forces. In this way the tensegrity structures are changed from resonant to rigid, subsequently consolidating into geodesic structures.

Fig. 1. Model of a rigid tensegrity icosahedron in bamboo

Observing a deresonated tensegrity structure, we are led to the conclusion that the underlying rods support those above, and that the structure works in compression as happens in traditional structures in which the bolts serve to prevent lateral sliding, thus giving rise to a mistaken idea of the dynamics of the system. In reality, the structure is pushed towards the exterior by a hidden tension system that recalls the latent explosion of a soap bubble, with the difference that in these structures the superficial external membrane is supported by tension forces that derive from the membrane itself.

The tensegrity universe of R.B. Fuller: Brief chronology

1927, Greenwich Village Studio. The first experiments of Fuller in the search for an integrated architecture in tension (tensegrity).

1949. From the collaboration with sculptor Kenneth Snelson (at that time a student of Fuller's), are born the first models of tensegity columns.

1953, Minnesota University. Realization at large scale of a rigid tensegrity structure with 270 non-identical rods. The structure would be patented in 1962. The patent covered tensegrity structures with identical rods as well.

1959, Oregon University. The first model of a deresonated tensegrity with 270 identical rods. The 270-rod structure was described on p. 394 of *Synergetics 1* as "isotropic tensegrity geodesic sphere: single bonded turbo triangles, forming a complex six frequency triacontahedron tensegrity."

1960, Long Beach State College. The realization of a rigid tensegrity structure in bamboo with a diameter of 14m. The outer ends of each element were calibrated with regards to the central point of the adjacent elements.

1961, Bengal College of Engineering, Calcutta. Bamboo dome, hypothesis of low-cost shelters that could be realized with local materials and technology. The bamboo dome was reproduced in *Domebook 2* and corresponds to the alternate breakdown of a 4v icosahedron, sectioned to 5/8 of a sphere.

1961, Southern Illinois University. Basket weave tensegrity. An interwoven dome of a 22m diameter, ¾ of a sphere, approximately 15m high. The struts were made of wooden centine, segmented and interwoven, with permitted the cost to be reduced to 1/5 of the equivalent traditional construction. This geodesic also corresponds to the alternate breakdown of a 4v icosahedron, sectioned to 5/8 of a sphere.

John Warren and Norman Foster

John Warren was born in 1946 in Corona del Mar, California. He studied art and biology, but preferred sculpture to all other activities. Today he lives in the Kaui islands of Hawaii. He worked with Fuller from 1971 on, realizing the models of the *Fly Domes* as well as the rigid tensegrity structures (deresonated tensegrities) having the geometry of the alternate breakdown of a 6v icosahedron sectioned to 5/8 of a sphere. In 1982-3 he collaborated with Norman Foster & Associates, realizing the models of the Autonomous House. The Autonomous House was to be the residence of the Fullers in Los Angeles. Partly transparent and partly opaque, the dome was fitted with two rotating spherical caps that could be darkened and were able to follow the course of the sun during the day. The interstices between the two domes enclosed hot or cold air to heat or cool the interior. Tubes in carbon fiber were envisioned.

1986. Windstar Biodome, Snowmass, Colorado. I described this dome on p. 15 of the magazine *Bioarchitettura* 18 (June 2000) and is related to the pioneering work of the New Alchemy Institute (1977), as well as to the concept of "permaculture" developed by Bill Mollison in 1978. The Biodome was realized on the model of the interwoven dome (basket weave), with the result that it is a rigid tensegrity structure. It is covered with inflatable three-ply plastic (EFTE) cushions that contribute to the greenhouse effect, optimizing the values of insulation, light transmission and durability (lasting about 12 years). The same panels were used in the Eden Project by Grimshaw and Partners in St. Austell in Cornwall (1988). The whole structure was assembled by hand. The reflective surface of the cushions reduced the nighttime heat loss to a minimum. Acquaculture was carried on in the interior of the Biodome, the water remaining tepid thanks to the constant production of passive solar heat.

The alternate breakdown of the iscosahedron

In alternate breakdown, faces of the iscosahedron are divided into multiple frequencies of 2. The frequency of subdivision was indicated by Fuller with "v," due to the similarity between the letter v and the triangle. After the great circles (GC, the equator that divides the sphere into two equal parts), next to be taken into consideration are the small circles (SC, all the circles of the sphere whose centers do not coincide with the center of the sphere). This distinction is necessary because the sphere of frequencies 4 and 6 present a degree of symmetry that is distorted and modified with respect to the 2v geodesic. The introduction of the SC allows the sectioned geodesic to rest on a single plane.

The icosahedron Alt 2v (fig. 2) has 6 decagonal GCs arranged so as to form a spherical icosadodecahedron with 12 pentagons and 20 triangles. A vertex of the pentagon meets the vertex of another pentagon. The circuit model has 30 struts that form 6 interwoven pentagonal circuits. The 60 tendons define the corners of the icosadodecahedron (figs. 3-6).

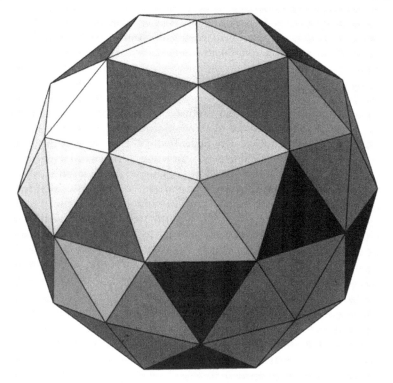

Fig. 2. Icosahedron Alt 2v geodesic sphere

Fig. 3. Model of a geodesic dome icosahedron Alt 2v, in bamboo

Fig. 4. Reciprocal model of the icosahedron Alt 2v in bamboo, built on sand

Fig. 5. Model of a reciprocal frame icosahedron Alt 2v, 5/8 of a sphere. The covering is made from a cloth hung on the interior

Fig. 6. Reciprocal frame in bamboo, icosahedron Alt 2v, 5/8 section of a sphere, diameter 2.92 m, maximum height 1.83 m. If covered, could be used a a shelter from the sun

The icosahedron Alt 4v, has 12 20-sided SCs. The SCs are arranged in parallel couples. Doubling the 6 GC of the preceding model gives rise to six parallel couples of circles, no longer great but small, because the GC is reduced in diameter as it gets further from the equator. The resulting polyhedron will have 12 regular pentagons, 30 hexagons and 80 irregular triangles. The geometric arrangement from vertex of pentagon to vertex of pentagon is penta/hexa/penta. The circuit model is made of 120 struts and 240 tendons arranged in 12 decagonal circuits (figs. 7-8).

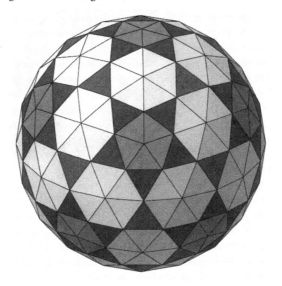

Fig. 7. Geodesic sphere icosahedron Alt 4v

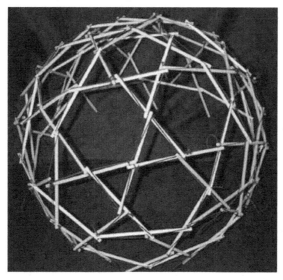

Fig. 8. "Circuit" model, tensegrity structure icosahedron Alt 4v

The icosahedron Alt 6v has 270 struts, 540 tie rods and 18 SCs of 30 sides each with two different types of circuits. The geometric arrangement from vertex of pentagon to vertex of pentagon is penta/hexa/hexa/penta. This polyhedron is obtained by adding 6 SCs to the preceding model (fig. 9).

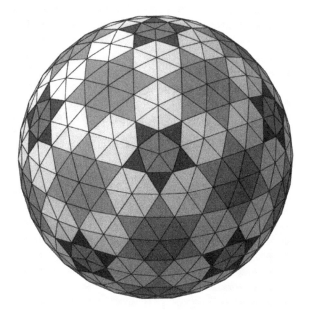

Fig. 9. Geodesic sphere icosahedron Alt 6v, often used as the reference geometry for Japanese thread balls known as *Temari*

Tensegrity structure Icosahedron Alt 4v, "circuit" model

The principle module is constituted of one strut (sliced strut) surrounded by a single or even a double tendon. The construction of the model begins with the upper pentagon, then five struts are added, followed by five more couples of struts and so on. It is practically impossible to visualize the figure without referring to drawings or to models.

For the construction of spherical models there are two principle references:

A) A first version with struts all equal and with the tendons about half as long as the struts results in a form that is less spherical and more angular. The struts wave in a way that is slightly disorderly.

B) A second, more spherical and more elegant, version utilizes struts with two different dimensions and three different types of tendons (which permits the visualization of the reference polyhedron). The dimensions of the tendons are the same utilized as geodesics in the struts of the bamboo dome. In this version [Pugh 1976: 62], 60 22.5 cm long (L) struts are used, and 60 20.31 cm short (S) struts are used, which alternated around the structure. The pentagon of departure is made of 5 struts and is visualized from the short tendons. The dimensions of the tendons are L= 11.87 cm; M (medium) = 11.25 cm; S = 9.67 cm. The tendons are arranged in a LMSML sequence.

Takraw Ball

The takraw ball is used in Malaysia and Thailand in a game called *Sepak Takraw*. The ball is created by interwearing six large circles, each of which is formed of from six to eleven strands of rattan. The modern version is made of plastic. The Thai ball (fig. 10) is based on frequencies equal to that of the iscosahedron. Observing this spherical geodesic it can be noted that the sides of the regular pentagon are smaller with respect to the sides of the irregular hexagon, in a ratio of about 2 to 3. It isn't easy to come across the data for the construction, because in Thailand this art is passed down orally, conserved in the memories of the basket masters. In the cardboard model of Iscosahedron Alt 4v that I made, to the side of the pentagon equal to 4 cm corresponds a height of the strip that equals 9 cm (by "strip" we mean one of the 6 GCs that make up the wicker ball).

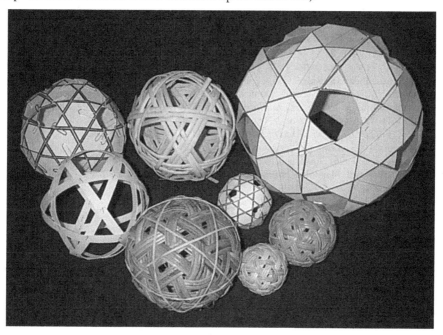

Fig. 10. Studies of the geodesic geometry of the Takraw ball.

Oriental culture is rich in other examples. In the Forbidden City of Beijing, in front of the bridge of Purity of Paradise, is the sculpture of the Lion holding in a geodesic ball in his front balls that dates from the Quian Long Dynasty (1736-1796), the geometry of which could refer to the alternate breakdown of the icosahedron at a frequency greater than 6. Some modules, however, turn out to be irregular. (Both Joseph Clinton and Russel Chu agree about the reference to the geodesic geometry.) It is supposed that the sphere derives from the art of the Temari, which consists of a ball made with strips of silk, used as a toy or as decoration in ancient China and in Japan.

The photograph on p. 8 of [Hargittai 1995] reveals the geometry of the truncated icosahedron (the fullerene) combined with the geometry of the icosahedron alt 6v. The alternate breakdown of the icosahedron 6v can be traced also in the geometrical arrangement of the pentagonal and hexagonal stars.

Conclusions

The reciprocal structures of Leonardo can be considered as forerunners to rigid tensegrity structures, which can themselves be considered the forerunners to geodesic structures. The rigid tensegrity systems turn out to be apparently more complex than reciprocal structures but in reality the only difference regards the systems of joints: in reciprocal structures the terminal point of a rod corresponds to the terminal point of another rod, and the final joint assumes the aspect of a "turbinated" star and is easily modified. In rigid tensegrity structures as well the final joint assume a "turbinated" aspect (as Fuller stated in his patent of tensegrity structures in 1962). Once again, today as in the past, the genius of Leonardo indicates new yet ancient solutions and points to the simplest possible way forward in research that is often apparently complex and confused.

Translated from Italian by Kim Williams

Bibliography

Bioarchitettura **18** (2000), **21** (2000), **23** (2001). Bolzano.
CHILTON, BOO, LEWIS, POPOWIC. 1977. Structural Morphology: Towards The New Millennium. *Dome Magazine*, Wheat Ridge, Colorado.
Domebook 1 & 2. 1970-1971. "Pacific Domes". Bolinas California
FULLER, R. Buckminster. 1963. *Ideas and Integrities*. New York: Macmillan.
———. 1975-1979. *Synergetics 1 & 2*. New York: Macmillan
———. 1973. *The Dymaxion World of Buckminster Fuller*. New York: Doubleday.
GOULD, Claudia. 1993. *Kawamata Project on Roosevelt Island*. Tokyo, 1993.
GUTDEUTSCH, Gotz. 1996. *Building in Wood*. Birkhauser.
HARGITTAI, Istvàn and Magdolna HARGITTAI. 1995. *Symmetry through the Eyes of a Chemist*. Plenum Press, 1995.
International Journal of Space Structures, vol. 17. 2 & 3, 2002.
KENNER, H. 1976. *Geodesic Math*. University of California Press.
L'Architettura Naturale, 10/2001. Milan.
MCHALE, John. 1964. *R. B. Fuller*. Il Saggiatore.
PEDRETTI, Carlo. 1978. *Leonardo Architetto*. Milan: Electa.
POPOVIC Olga Larsen, 2003. *Conceptual Structural Design: Bridging the Gap Between Architects and Engineers*. London: Thomas Telford.
PRENIS, John. 1973. *The Dome Builder's Handbook*. Running Press.
PUGH, A. 1976. *An Introduction To Tensegrity*. University Of California Press.
———. 1976. *Polyhedra: A Visual Approach*. University of California Press.
ROBBIN, Tony. 1996. *Engineering A New Architecture*. New Haven: Yale University Press.
WRENCH, Tony. 2001. *Building a Low Impact Roundhouse*. Hampshire: Permanent Publications.

About the author

Biagio Di Carlo is an architect as well as a graphic designer and a musician. He received his degree in architectural studies with honors in 1976 from the University of Architecture of Pescara; his thesis was published. His thesis advisors were Eduardo Vittoria, together with Giovanni Guazzo and Augusto Vitale. He taught at the Art Institute of Pescara, and often collaborates with the architectural faculties of Pescara and Ascoli, by giving lectures, lessons and seminars in synergetic geometry, geodesic domes, tensegrities, quasi crystals and four dimensional polytopes. He is the author of three self-published books: *Cupole geodetiche, Poliedri* and *Strutture Tensegrali*, and has published articles on synergetic geometry and geodesic domes in journals such as *Bioarchitettura* and *L'Architettura naturale*. Among his interests are architecture, molecular geometry, graphic arts and cartoon illustrations, Latin jazz, song writing and Brazilian music. More information about his activities is available on his website: http://www.biagiodicarlo.com.

Sylvie Duvernoy

Via Benozzo Gozzoli, 26
50124 Florence ITALY
sduvernoy@kimwilliamsbooks.com

Research

Leonardo and Theoretical Mathematics

Keywords: Leonardo da Vinci, Luca Pacioli, quadrature of the circle, doubling the cube, Renaissance mathematics

Abstract. Leonardo's mathematical notes bear witness to a work in progress and allow us to look directly into the mind of the writer. In Leonardo we find two of the three fundamental classical geometric problems: the duplication of the cube and the quadrature of the circle. While Leonardo is extremely familiar with two-dimensional geometry problems, and proposes playful graphic exercises of adding and subtracting polygonal surfaces of all kinds, he is still unable to solve the problem of the duplication of the cube. Numerous pages testify of the attempt to rise above planar geometry and reach the realm of the third dimension, but Leonardo always bumps against the limits of quantity calculation possibilities of his age.

Introduction

Leonardo came to the study of mathematics rather late in life. We know from Giorgio Vasari that he attended in his youth, as any student of his time, the *scuola d'abbaco*, where he is presumed to have learned the bases of arithmetic and geometry. Vasari also reports that he was so clever that he did not attend the school for more than a few months, and soon left because he used to argue with the teacher, who was not able to give satisfactory answers to his objections [Vasari 1991: 557-558].

We must believe the biographer when he tells us that Leonardo quickly decided to leave the school, but we cannot agree with him when he says that he did so because he was too good a student and would not increase his knowledge in this popular institution. Until he met Luca Pacioli in Milan, whom he accepted as a teacher and a master, Leonardo was far from being a brilliant mathematician. Looking through the various folios of the early codices we can see that he was unfamiliar with arithmetic, and very clumsy in computational operations involving fractions, both in multiplication and division. He would not believe, for instance, that the division of a number (or fraction) by a number (or fraction) inferior to one would give a result superior to the original number. He also used to make basic mistakes when multiplying large numbers including zeros.

The zero symbol, 0, had been introduced into the Western arithmetic annotation system, together with the full series of Arab numerals – 1,2,3, ... 9 – by Leonardo of Pisa, better known as Leonardo Fibonacci, nearly three full centuries earlier, in 1202, with the publication of his famous *Liber Abaci* which at the time of Leonardo da Vinci's youth was still the main textbook on which teachers relied for their lessons.

Leonardo da Vinci and Luca Pacioli met in Milan in 1496, at the Court of Ludovico Sforza. Leonardo had been in the Duke's service since 1482, and he was 44 years old when he first met Luca Pacioli, who had been called by the Duke to teach mathematics. Pacioli himself was 51, and had just published two years before, in 1494, his *Summa de aritmetica geometria proportioni e proportionalità*, which was for the most part a revisitation of the *Trattato d'abaco* written (but not published) by his own master, Piero della Francesca.

The encounter with Pacioli marks a turning point in Leonardo's life as regards the study of mathematics. Guided by his master and friend, he started a systematic study of theoretical mathematics, going from recent and contemporary publications back to classical sources and textbooks. It is clear that, just like every scientist of his period, he carefully studied the *Elements* of Euclid, and became familiar with all the classical geometrical problems.

We have at our disposal a fair number of architectural and mathematical treatises from the Renaissance period, but the preliminary research notes necessary for the compilation of those books were all lost. We only have the final compositions, printed and illustrated in order to offer a didactic edition. Leonardo, on the other hand, never wrote a proper treatise, but left to posterity a huge quantity of manuscript papers that can be considered as preliminary notes for books never written. His notes, although confused and somewhat disordered, are very precious to us because they testify to work in progress and allow us to look directly into the mind of the scientist. While real treatises only show the solutions to problems and the certified rules, in Leonardo's manuscripts we find numerous questions that sometimes reach a conclusion, sometimes not, giving us valuable information as regards the process of mathematical research in the Renaissance period, covering a wide range of approaches, from graphic and arithmetic, to geometric and analytical.

The duplication of the cube

The first striking thing to notice is how at least two of the three fundamental classical geometric problems were still present in the minds of the scientists of Renaissance times: the duplication of the cube and the squaring of the circle.

Leonardo put a lot of effort into trying to solve the problem of the duplication of the cube. This problem, according to the legend attached to it, is the most ancient example of a relationship between architecture and mathematics. The people of Delos were faced with an architectural conundrum concerning a religious monument. The oracle had told them to build an altar to Apollo twice as big as the previous one, which was of a cubic form. But what should the dimension of the side of the new cube be in order to obtain a cube twice the volume of the original one? After a first, wrong, attempt consisting in doubling the side of the cube, the architectural request was for mathematicians, who had not yet discovered the algebraic calculation of irrational quantities. Leonardo's predilection for this specific problem surely comes from his obvious interest in three-dimensional geometry and stereometry. His many efforts to solve the problem go from somewhat ingenuous and empirical attempts, to the study of classical solutions, and follow different sorts of scientific research methodologies.

A graphic approach: *Codex Arundel*, folio 283v. Leonardo asks himself whether any kind of simple extension from two- to three-dimensional geometry exists (fig. 1).

Can Plato's theorem on the duplication of the square be extended to the duplication of the cube?

Is the volume of a cube built from a double square twice the volume of a single unity cube?

If the diagonal (or diameter) of a square with a side of 1 is the graphic visualisation of the incommensurable quantity of the square root of two, is the diagonal of a cube with a side of 1 the graphic answer to the irrational number equal to the cube root of 2?

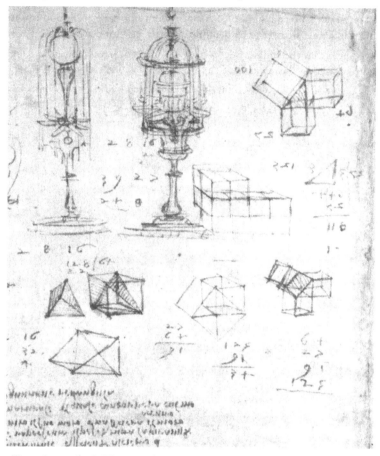

Fig. 1. Leonardo da Vinci, *Codex Arundel*, fol. 283v. Graphic and arithmetical approaches to the duplication of the cube, starting from the theorems of Plato and Pythagoras

The answers are no. The diagonal of the cube is equal to the square root of 3, and not the cube root of 2, which is a smaller number than the square root of 2.

An arithmetic approach: *Codex Arundel*, folio 283v. Following another idea, Leonardo then tries to extend Pythagoras's theorem on right-angled triangles from squares to cubes. If the sum of the squares of the sides of these triangles is equal to the square of the hypotenuse, can the same apply to cubes?

Taking the simplest example, the 3-4-5 triangle (the so-called "Egyptian triangle"), the calculation quickly appears disappointing.

More generally, is there a cubic number that can be split into the sum of two lesser cubic numbers? Would those three numbers lead to the discovery of a particular family of triangles?

The answer is no. The equation $a^n + b^n = c^n$ does not have a solution in integers for $n > 2$. But this theorem had yet to be demonstrated. It was not before 1753 that Leonhard Euler demonstrated that the equation $a^3 + b^3 = c^3$ does not have a solution. And the final demonstration of the so-called "Fermat's third theorem", which is that the general equation $a^n + b^n = c^n$ does not have a solution for $n > 2$, was given by Andrew Weil in 1993.

A stereometric approach: *Codex Arundel* folios 223v and 223r.

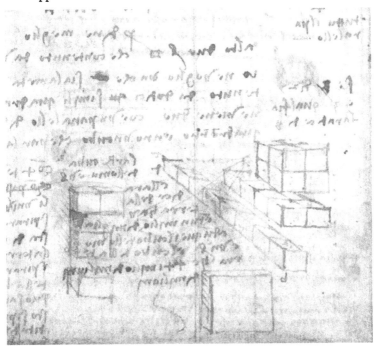

Fig. 2. Leonardo da Vinci, *Codex Arundel*, fol. 223 v. Stereometric approach to the duplication of the cube

Other pages testify to a strenuous effort to solve the problem of Delos according to a stereometric approach. Instead of doubling a cube Leonardo reverses the problem and tries to divide one cube into two smaller and equal ones. He starts with a cube that he divides into 27 small units – which is easy because it was "made from the cube root of 27", equal to 3 – but 27 is an odd number whose units cannot be rearranged into two equal small cubes. So, Leonardo takes another cube made of 8 small cubic units and tries to work out how to arrange four of these units in a cubic form. In the meantime, he tries to find out if there is a direct proportional relationship between the surface and the volume of a solid. Is the envelope of a solid proportional to its volume? This eventuality could be a convenient solution to bring back the problem from three-dimensional geometry to two-dimensional, but he quickly understands that the idea is erroneous. The negative conclusion comes on the back of the same folio, which contains the affirmation "...so don't use this science of cubes according to their surfaces but according to their bodies, because a same quantity has different surfaces of infinite values... equal surfaces don't always contain equal bodies..."

So going back to the 8-unit cube, Leonardo wonders: " if I have a cube made of eight cubes, I made it from the cube root of eight... if the cube root of 27 is 3, which is the root of 8?"

Another arithmetic approach: *Codex Atlanticus,* folio 161r. Finding an approximate value for the cubic root of 8 would not have solved the problem of the duplication of the cube anyway, nor the reverse corollary, its division into two. Only the determination of an approximate value of the cubic root of 2 would have given an arithmetical solution to this mathematical problem (fig. 3).

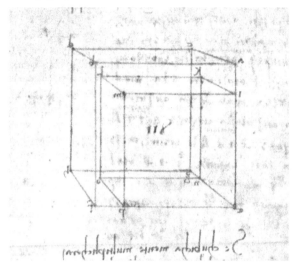

Fig. 3. Leonardo da Vinci,*Codex Atlanticus,* folio 161r. Arithmetical approximation: the value of cube root of 2 is close to 5/4

In the published mathematics books of Leonardo's times (*Summa de aritmetica geometria proportioni e proportionalità* by Luca Pacioli, etc...), whereas the value of π had long been considered almost equal to 22/7, and square root of 2 nearly equivalent to 7/5 (or 14/10), the cubic roots, which are not equal to a round number, such as the cube root of 27 or 64, are not approximated by a fraction, but remain as "cubic root of *x*" and the authors do not give estimated values for them.

Leonardo reached an acceptable approximation for the cube root of 2. Successive calculation attempts lead him to conclude that a cube of a 5-unit side has a volume close to twice that of a cube of a 4-unit side. 125/64 is accepted as a good approximation of 128/64 = 2.

5/4 can therefore be considered a close approximation of the cube root of 2, which can consequently be adopted from then on for the practical purposes of three-dimensional metrical geometry (fig. 4).

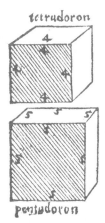

tetradoron

pentadoron

Fig. 4. Evidence of ancient approximation to the irrational value of cubic root of 2:
the cubic Greek bricks of sides 4 and 5 (*tetradouron* and *pentadouron*) mentioned by
Vitruvius in *De Architectura*, bk. II, chap. 3, drawn by Andrea Palladio for Daniele
Barbaro's Renaissance translation and commentary of Vitruvius (1556)

A classical geometric approach: *Codex Forster* I, folio 32. In addition to looking for his
own solutions to the duplication of the cube, Leonardo also studied the classical solutions
of the ancient Greek mathematicians, probably guided by Luca Pacioli, who was a scholar
of Euclid. Evidence of this can be seen in the carefully drawn interpretation of Apollonius's
method for the Delian problem (fig. 5).

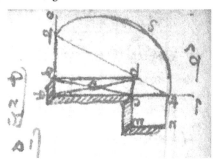

Fig. 5. Leonardo da Vinci, *Codex Forster* I, folio 32. A classical geometrical approach to the
duplication of the cube, from Apollonius' method

Hippocrates of Chios had reduced the problem of the duplication of the cube to the
problem of finding two mean proportionals between two straight lines representing two
arithmetical magnitudes.

The three most famous answers to the query are the work of three mathematicians of
the Platonic era: Archytas, Eudoxus and Menaechmus. These solutions were followed by
several others, including one attributed to Apollonius that is particularly simple both
conceptually and graphically. Apollonius's method is not among the two classic solutions
that Vitruvius mentions in his treatise, and that Barbaro was to illustrate in his Renaissance
commentary of Vitruvius, adding Nicomede's proposal. This method is derived from
Euclid, and more precisely from Book 2, last proposition: from a given rectangle, find an

equivalent square; or on the other way round: from a given square, find the equivalent rectangle having a given base (fig. 6).

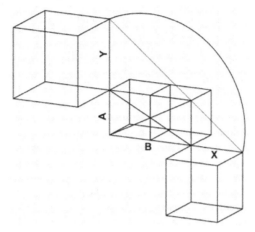

Fig. 6. Apollonius's solution to the duplication of the cube. Drawing by the author

The demonstration runs as follows: if the two initial straight lines (A and B) are assembled to form two adjacent sides of a rectangle, a ruler must be placed on the opposite vertex of that parallelogram, and swung around the pivot thus formed until it bisects two lines extending out from the initial sides of the rectangle at two points that are equidistant from the rectangle's center. The equidistance is verified – and demonstrated – by drawing an arc traced with a compass whose needle is placed at the center of the rectangle: i.e., the intersection point of its diagonals. The values of the two intervals thus obtained (X and Y) between the sides of the rectangle and the intersection points will be the two sought-after mean proportionals (fig. 7).

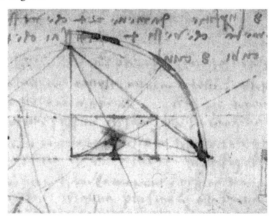

Fig. 7. Leonardo da Vinci, *Codex Arundel*, f. 223v. Leonardo's copy of a classical two-dimensional graphic explaining Apollonius's solution

In accordance to the Greek tradition of mathematical sketches, the diagram drawn by Apollonius to illustrate his demonstration is extremely schematic and two-dimensional: it represents the partial orthogonal projection of volumes on a plane parallel to one of their faces. It can be considered to be either a top view – *ichnographia* – or a front view – *orthographia*. The rectangle seen in the diagram is the face of a parallelepiped whose depth is equal to its width: a prism with a square base (invisible on the figure because of its perpendicularity to the drawing plane), and a given height. Apollonius's demonstration can be understood with the help of the figure only if the reader of the image is able to interpret it correctly by substituting the missing information regarding the third dimension with a mental procedure that will complete the message. Leonardo, while studying and sketching, added the third dimension, transforming the figure in a perspective (axonometric?) drawing in order to ease comprehension (see fig. 5).

Leonardo's drawing shown in *Codex Atlanticus* fol. 588 r (fig. 8) illustrates the case in which the cylinder is a double cube, and makes a discovery when noticing (probably by chance) that BF is equal to BE. This implies that the geometric construction can be reduced to a very quick and easy manipulation of the single straightedge (with the very slight support of a compass), and represents an important step towards the simplification of the solving of this problem, which has inspired the most sophisticated inventions and construction of heavy mechanical drawing tools since antiquity. Leonardo's method makes it possible to skip Apollonius's mechanical test of the simultaneous line balancing on point A, and arc drawing with center in M, which Leonardo judged to be imprecise and dubious, with a result that can only be obtained through *faticoso negozio*, "tiring effort".

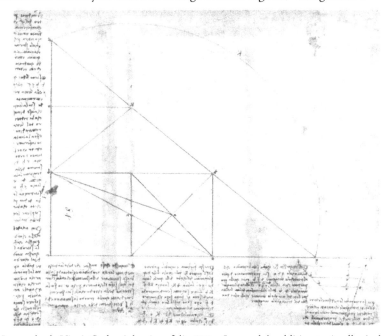

Fig. 8. Leonardo da Vinci, *Codex Atlanticus*, folio 588 r. Leonardo's addition to Apollonius's solution

Simplification means divulgation and popularization. The graphic representation of this unknown and incommensurable quantity of the cube root of 2 has become as easy as – for instance – the construction of a pentagon inscribed in a circle. But Leonardo admits to not being able to draw an explanatory theory from his own finding, and this is why we may suppose that he reached it in a totally empirical way. Scientific theoretical demonstration is not Leonardo's specialty. Investigation and experimentation tend to stop when a discovery is made, in hope and haste to make another. The caption next to the figure in *Codex Atlanticus* shown in fig. 8 says: "If you will tell me for what reason the half diameter of the circle fits six times into its circumference and why the diagonal of the square is not commensurable to its side, I shall tell you why the straight line that goes from the upper vertex of one of the two joined squares to the center of the second square shows us the cube root of the two cubes reduced in a single one".

The squaring of the circle

Leonardo also spent a lot of time trying to solve a second classical problem: the squaring of the circle. One day he even claims to have reached a solution: on *Codex Madrid* II, folio 112r we read, "the night of St Andrew, I finally found the quadrature of the circle; and as the light of the candle and the night and the paper on which I wrote were coming to an end, it was completed; at the end of the hour." But the solution is not there…

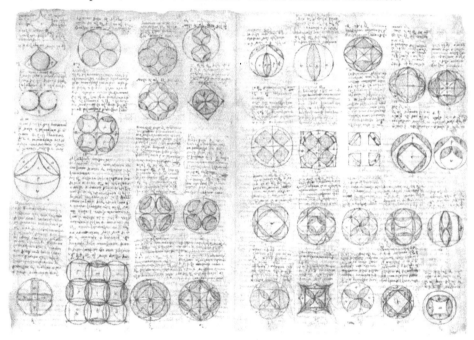

Fig. 9. Leonardo da Vinci, *Codex Atlanticus*, folio 471. Squaring the circle, graphic research

Leonardo's approach to the attempt of squaring the circle is obviously inspired by that of Archimedes, even if it is not clear whether it is by direct reading or only by second-hand knowledge. In any case, he remained unsatisfied with the approximate ratio between the circumference and the diameter as 22/7. Therefore he tries to take this approximation beyond the 96-sided polygon, in an attempt to bring the difference of areas between circle and polygon to be as small as the "mathematical point", which has no quantity. This research generates an enormous quantity of sketches that show an infinite variety of decorative shapes (*Codex Atlanticus*, fol. 471, fig. 9). Scientific research turns into a never-ending, playful,l graphic game. Leonardo intended to write and publish a treatise in order to make public his discovery, and its title would have been *De Ludo Geometrico*. This methodology does not lead to a satisfactory solution, or even to any progress towards a result, but the value of his effort lies in this attempt at extension ad infinitum.

Leonardo's contribution to mathematic research

It is in the realm of three-dimensional geometry that Leonardo achieved his greatest result: the determination of the location of the center of gravity of a pyramid.

A mechanical approach: codex Arundel, folio 218 v. The elevation from two- to three-dimensional geometry starts with the study of Archimedes' book, *On the equilibrium of planes*. Leonardo must have felt at ease with Archimedes' experimental method, where the planes are considered to have a weight and are hung at the end of levers and ropes in order to determine the exact position of their center of gravity. Archimedes deals with planes, especially triangles, while Leonardo extends the experiment to solids, and first of all to the regular tetrahedron. Knowing from previous studies the position of the centers of gravity of the faces of the solid, he finds out that "the center of gravity of the body of four triangular bases is located at the intersection of its axes and it will be in the 1/4 part of their length" (fig. 10).

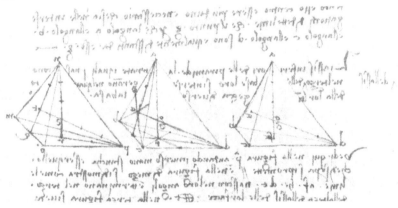

Fig. 10. Leonardo da Vinci, *Codex Arundel*, fol. 218v. The centre of gravity of a pyramid

The generalization of this discovery leads to the statement that "the center of gravity of any pyramid – round, triangular, square, or of any number of sides – is in the fourth part of its axis near the base."

On *Codex Arundel* folio 123v there is an additional theorem concerning the tetrahedron:

> the pyramid with triangular base has the center of its natural gravity in the segment which extends from the middle of the base [that is the midpoint of one edge] to the middle of the side [that is, edge] opposite the base; and it is located on the segment equally distant of the line joining the base with the aforesaid side.

Conclusion

From this brief study of the works of Leonardo in the area of theoretical mathematics it appears that stereometry and solid geometry were the fields that best suited his inventive skills, and this is probably due to his skill in three-dimensional representation, which allowed him to obtain an exact visualization of the objects of his studies. All the folios of the various codices are full of perspective sketches that are not drawn in compliance to the recently established *costruzione legittima*, but rather following a spontaneous gift for representation that often generates some kind of pre-axonometric drawings rather than perspective ones.

On a more general level, we may conclude that Leonardo contributed to mathematical and scientific research in the Renaissance period by demonstrating the power of the tool of three-dimensional representation as a research device as well as a persuasive instrument. The well-known series of drawings of the Platonic – and non – solids that he made as illustrations for the book of his friend Luca Pacioli is simply one of the many examples of this.

Bibliographic references

ARTMANN, Benno. 1999. *Euclid – The creation of mathematics.* New York: Springer-Verlag.
BAGNI, Giorgio T. and Bruno D'AMORE. 2006. *Leonardo e la matematica.* Florence : Giunti.
EUCLID. 1993. *Œuvres.* Ed. and trans. Jean Itard. Blanchard, Paris
FOWLER, David. 1999. *The mathematics of Plato's academy. A new reconstruction.* Oxford: Clarendon Press,.
HEATH, Sir Thomas. 1981. *A history of Greek mathematics.* New York: Dover Publications.
KNORR, Wilbur Richard. 1986. *The ancient tradition of geometric problems.* Boston: Birkhäuser.
LORIA, Gino. 1914. *Le scienze estate nell'antica Grecia.* Milan: Hoepli.
MARCH, Lionel. 1998. *Architectonics of Humanism.* London: Academy editions.
MARINONI, Augusto. 1982. *La matematica di Leonardo da Vinci.* Arcadia.
PACIOLI, Luca. 2004. *De Divina Proportione,* Augusto Marinoni ed. Milan: Silvana editore.
TATON, René, ed. 1958. *Histoire générale des sciences.* Paris: Presses Universitaires de France.
VASARI, Giorgio. 1991. *Le vite dei più eccellenti pittori, scultori e architetti* (1550). Rome: Newton Compton editori.
WITTKOWER, Rudolf. 1962. *Architectural principles in the Age of Humanism.* London: Academy editions.

About the author

Sylvie Duvernoy is an architect who graduated from Paris University in 1982. She participated in the Ph.D. program of the Architecture School of University of Florence and was awarded the Italian degree of "Dottore di Ricerca" in 1998. She presently teaches architectural drawing at the engineering and architecture faculties of University of Florence. Her research has mainly focused on the reciprocal influences between graphic mathematics and architecture. These relationships have always been expressed by the means of the drawing: the major and indispensable tool of the design process. The results of her studies were published and communicated in several international meetings and reviews. In June 2002 she presented "Architecture and Mathematics in Roman Amphitheaters" at the Nexus 2002 conference in Óbidos, Portugal. In addition to research and teaching, she maintains a private practice as an architect. After having worked for a few years in the Parisian office of an international Swiss architecture firm, she is now partner of an associate office in Florence, the design projects of which cover a wide range of design problems, from remodelling and restoration to new constructions, in Italy and abroad.

Mark Reynolds

667 Miller Avenue
Mill Valley, CA 94941 USA
marart@pacbell.net

Keywords: Leonardo da
Vinci, octagons, geometric
constructions

Research

The Octagon in Leonardo's Drawings

Abstract. Mark Reynolds presents a study on Leonardo's abundant use of the octagon in his drawings and architectural renderings. Specifically, he focuses on Leonardo's applications of the octagon: in his studies and sketches of the centralized church, and for which we can find influences specifically from Brunelleschi, as well as from other fifteenth-century architects working with this type of religious structure; in his almost obsessive and frequently repetitious drawing of octagonal shapes and forms in his notebooks throughout his career; in his project for a pavilion while with the Sforzas in the last part of his period in Milan. Also examined are ways to develop the modules to accommodate $\sqrt{2}$ and the θ rectangles. The application of the modular units, so far, have been within the square and its gridwork, but as the octagon has traditionally been used in the development of both the circle and the square, this shape is an interesting challenge in terms of linking the two-dimensional surface to the three-dimensional forms we are planning to generate. The object is to provide us with more insight as to why the octagon held so much fascination for Leonardo as one of the ultimate geometric expressions of grandeur and practicality in spatial organization, design, and development.

Often in Leonardo's drawings of octagons, precise geometric constructions were lacking; the master's approach was freehand. The author seeks to learn if Leonardo's sketches can be put to the rigors of strict geometric construction, and still be viable as accurate renderings of octagonal geometric spaces with his own geometric constructions of those same spaces.

Geometry existed before the creation of things, as eternal as the Spirit of God; it is God Himself and He gave Him the prototypes for the creation of the world.

Johannes Kepler
Harmonices Mundi, 1619

Although God Himself delights in the odd number of the Trinity, nonetheless He unfolds Himself profoundly through the quadrinity in all things."

Giordano Bruno
On the Monas, ca.1598
Hamburg Edition, 1991

I Introduction

It is obvious to anyone spending time looking through Leonardo's sketchbooks that he drew octagons and octagonal systems almost obsessively throughout his long artistic career (fig. 1).[1] Even when he did not draw the completed octagon, we still find Leonardo's preoccupation with octagonal geometry. In rapid sketches and in considered drawings, from floor plans to mechanical devices, we find octagonal shapes taking precedence over other geometric systems. At other times, Leonardo's interest in the square and its diagonal, requisites for √2 geometry and the octagon, can also be found scattered throughout his studies. One well-known example is in the *Codex Atlanticus* (Cod. Atl), fols. 190 v-b. It shows the head of a warrior for *The Battle of Anghiari*, with notes on geometry that include several square and diagonal constructions. There is a cube with a diagonal on one of the square faces. The cube also has an internal diagonal that runs inside the form, from front corner to rear corner of the cube, which demonstrates the √2 and √3 lengths within the form.

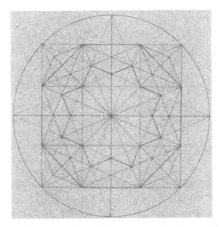

Fig. 1. Octagon and octagonal star, *Cod. Atl.* fol. 223 r-a, after Leonardo; pen and ink and conte crayon on ochre paper. All figures are by the author, after Leonardo.

The construction demonstrates that Leonardo's explorations were far more than rudimentary. A drawing for the plan of the city of Imola, in 1502 (Windsor, RL 12284) shows a plan view of the city drawn in a circle divided into eight parts (with four subdivisions of each of the eight sections). Several drawings of octagon-based fortifications done in 1504 can be found in *Cod. Atl.* Fol. 48, v-a. *Cod. Atl.* Fol. 286 r-a, of technological studies and wooden architecture (an "anatomy theatre"?), shows a circle divided into eight parts, each containing its own circle. There is also a famous sheet of sketches for the *Last Supper* and geometrical drawings in the Royal Library (Windsor RL, 12542).

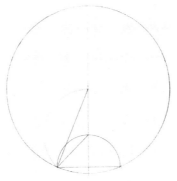

Fig. 2. Construction for an octagon inscribed in a circle, Windsor 12542, .after Leonardo; silverpoint

Predominating the middle of the page is a circle, drawn with compasses. At its base, we find the construction for generating an octagon (fig. 2) that would fill this circle, but Leonardo was satisfied not to complete the polygon within that circle. Perhaps he saw it completed in his own mind and found no need to draw it out, or perhaps this was his way of showing a most essential part of the octagon: how it can be generated within the circle from the diagonal of a generating square. It is striking to see the rapid studies for Christ and His Apostles juxtaposed with this exacting geometric construction. There are also other, smaller octagons on the page. This list of octagons could go on for quite some time, but for us the question is why Leonardo seemed to favor this shape over other geometric shapes, and why he investigated octagonal systems above all others as the basis for centrally-planned architecture.

Leonardo used the octagon in his architectural drawings almost to the exclusion of any other polygon, save a rare foray into a hexagon or an occasional dodecagon.[2] There were barely any architectural investigations done with the pentagonal system, other than a couple of fortresses and defensive constructions. There are times when Leonardo explores hexagonal geometry, occasionally in a temple plan, but more often when working with tessellations and patterns, including knot designs and embroidery. We do also find drawings relating to the circle and the square. Yet the octagon is related to these two shapes, and so, we are left with the predominance of the octagon and Leonardo's relentless curiosity about it. But why choose the octagon over the other regular polygons when he knew of them all, how to draw them, their armatures and proportioning systems? Perhaps we can find a few possible answers from looking at the time in which he lived, and the prevailing state of the world of art, architecture, and geometry into which he was introduced and that he developed.

II On the Nature of the Octagon

In art schools, architecture and design students are required to have two right-angled triangles, traditionally known as "set squares." They are the $45°/45°/90°$ triangle and the $30°/60°/90°$ triangle. For centuries, these tools have been traditionally used for *ad quadratum* ("from the square," $45°/45°/90°$) and *ad triangulum* ("from the triangle," $30°/60°/90°$) geometry. The two systems of proportioning were learned early in one's training in geometric construction, and remained part of the geometer's toolbox throughout one's career. Although a somewhat broad generalization, we may say that we usually find the development of ad triangulum in Eastern cultures, Byzantium, and in the early part of the European Middle Ages, while ad quadratum seems to have found more of a home in ancient Greece and Rome [Watts and Watts 1986; Watts 1996]. The square, $\sqrt{2}$ geometry, and the θ rectangle ($\theta = \sqrt{2}+1$) continued to be developed and utilized in European countries well into the Renaissance. (The subcontinent and China had always had an interest in the application of ad quadratum geometry as well.) The square, both as a system in and of itself, and as an integral part of all rectangles, was commonplace, and frequently, the circle and the square could be found in various combinations with each other and employed in various rectangular ratios. But it was Leonardo's ability to combine these other shapes with the octagon, and how these earlier uses of octagonal systems and symmetries may have influenced his own work that is of particular note. It is with this backdrop of ad quadratum geometry in Europe then that we begin our discussion of the octagon.

In the Quattrocento, geometry, symbolism, and the spiritual were still closely bound; studies on this subject are sometimes referred to as philosophical, sacred, or contemplative geometry. In this field of inquiry, the octagon was seen symbolically as the "intermediary" – the connecting shape – between the circle and the square [Cirlot 1962, 279]. The three shapes were traditionally drawn vertically, with the circle at the top representing the cosmos, and the square placed at the bottom to represent the earth. The octagon was drawn in between as that shape that connects the two. Occasionally, the octagon was viewed as a symbol for infinity. It was suggested that the octagon is a circle attempting to become a square, and a square attempting to become a circle. This concept was fully realized in the development of the Hindu vasta-purusha mandala, an ancient type of architecture [Stierlin n.d.: 43-58]. Mandala constructions combining the circle and the square usually result in the predominance of four or eight side figures. In China, the octagon represented a complex series of references, but we can say that its meaning was chiefly that of warding off evil entities and being a vehicle for the advent of good health and good fortune. In the number symbolism of Medieval Europe, eight was seen as representing cosmic balance and eternal life.[3] Related to this, the octagon also had deep significance for the Roman Catholic Church. The octagon and the star octagram were religious symbols for rebirth and resurrection. It was used in baptismal fonts in many churches, large and small. J.C. Cooper states that baptismal fonts were octagonal because the octagon symbolizes renewal, rebirth, regeneration, and transition [1978]. As Jay Kappraff has pointed out, "the star octagon, an ecclesiastical emblem, signifies resurrection. In medieval number symbolism, eight signified cosmic equilibrium and immortality" [Kappraff 2003, 127]. But were these more religious and metaphysical issues of interest to Leonardo? This symbolism may be tied into the centrally-designed church that utilized octagonal symmetry, but it is difficult to say with complete confidence that Leonardo used the octagon for these reasons. Yet, it is certainly possible.

Turning our attention to the mathematical qualities of the octagon, we find that Hermann Weyl, the great twentieth-century Princeton mathematician, made an important observation about the types of centrally-designed churches that Leonardo was designing. In his book *Symmetry*, Weyl writes,

> In architecture, the symmetry of 4 prevails… Central buildings with a symmetry of 6 are much less frequent… The first pure central building after antiquity, S. Maria degli Angeli in Florence, is an octagon. Pentagons are very rare [Weyl 1952, 66].

The octagon, by extension of the square, is then a structure that provided the Renaissance builder with one of the best geometric solutions for centrally designed ecclesiastical architecture where harmony could be maintained between the center and the immediate surrounding plan elements. Weyl continues:

> Leonardo da Vinci engaged in systematically determining the possible symmetries of a central building and how to attach chapels and niches without destroying the symmetry of the nucleus. In abstract modern terminology, his result is essentially our table of the possible finite groups of rotations (proper and improper) in two dimensions [Weyl 1952, 66] (fig. 3).

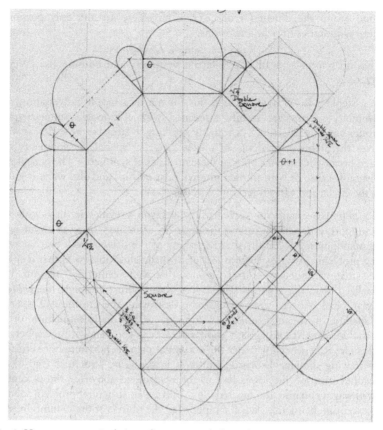

Fig. 3. Homogenous articulation of an octagonal plan, after Leonardo, *Cod. Atl.*, fol. 37 r-a; graphite on paper

In other words, the octagon gives the architect a broad range of spatial developments around a central design. And as we can see in many of Leonardo's octagonal drawings, he experimented widely with the many possibilities afforded by this shape. Within the limits of the two possible symmetry operations, Leonardo was able to come up with a great many variations, each with its own characteristics that made it unique within the general group.

Leonardo himself had suggested at one point that his geometric constructions were to be seen as "geometrical recreations", but this could have been just a rare moment of lighthearted sarcasm on his part, for Weyl goes on to discuss "Leonardo's Theorem." Leonardo devised a theorem on symmetry in which it is stated that,

> ...If improper rotations are taken into consideration, we have the two following possibilities for finite groups of rotations around a center O in plane geometry, which correspond to the two possibilities we encountered for ornamental symmetry on a line: (1) the group consisting of the repetitions of a single proper rotation by an aliquot part $\alpha = 360°/n$ of $360°$; (2) the group of these rotations combined with the reflections in n axes forming angles of $\frac{1}{2}\alpha$. The first group is called the cyclic group, C_n, and the

second group the dihedral group, D_n. Thus these are the only possible central symmetries in two-dimensions:

$$1. \quad C_1, C_2, C_3, \dots; \ D_1, D_2, D_3, \dots$$

C_1 has no symmetry at all, D_1, bilateral symmetry and nothing else [Weyl 1952, 65].

Professor Stephen R. Wassell clarified this for me in a recent conversation. He said, "Cyclic symmetry groups are readily associated with rotational symmetries, whereas dihedral groups contain reflections, or roughly speaking, bilateral symmetries, as well." The octagon gives the architect or designer a wide range of spatial developments around a central design because it contains both of these qualities of symmetry. The octagon is a very flexible polygon because of both its own internal relationships and also when it joined with the square, the circle, and various rectangles in the plane.

We also cannot overlook the fact that the octagon provides its user with a perfect vehicle for support, especially regarding the issue of "centering" from an architectural and engineering standpoint. Specifically, I am speaking of the octagon as a bracing system for load bearing problems. An important aspect of Brunelleschi's dome was that it spanned an immense distance without any central support, that is, a "centering" column. Leonardo certainly realized that the column would have been thrust through the center of an octagonal plane had it been necessary, thereby destroying the greatness of the span. An important quality of the octagon is its strength within its combination of vertical, horizontal and 45° diagonal armatures, and clearly, it can compete with the spherical dome as well. This fact is beneficial whether the octagonal plane is horizontal or vertical. I have included a drawing showing Leonardo's interest in centering structures (fig. 4). Often, Leonardo would truncate the lower half of the octagon to allow for the placement of a bridge or walkway. Leonardo also favored a 45° buttress in structural problem solving, with an obvious exception being the plans for supporting the tiburio of the Duomo in Milan.

Fig. 4. Centering drawing, after Leonardo, *Cod. Atl.*, fol. 200 r-a; pen and ink on red prepared paper

So then, evidence indicates that Leonardo worked with the octagon because it is both structurally sound and offers a broad range of design possibilities. From both utilitarian and aesthetic viewpoints, Leonardo realized that the octagon provided a great service in the area of architectural design, one that he chose to explore and develop throughout his life.

III Leonardo and the Octagon

Octonarius primus in numeris cubicis est, aeternae beatiudinus
nobis in anima et corpore stabilitatem simul at soliditatem designans
(Eight, as the first perfect cube [2^3], imprints us in body
and soul with the security of eternal beatitude)

Gernot of Rechersberg
Medieval German Theologian

When Leonardo went to Florence to live with his father and secure an apprenticeship in Verrocchio's studio, Santa Maria del Fiore was the talk of Europe. Filippo Brunelleschi had recently completed the dome of Florence's cathedral, and Leonardo's early exposure to the octagonal geometry of this structure (he would have been instructed in √2 and "sacred cut" constructions that yield the octagon) and in the Baptistery of San Giovanni are frequently cited as influences that left profound impressions on the young artist (figs. 5, 6).

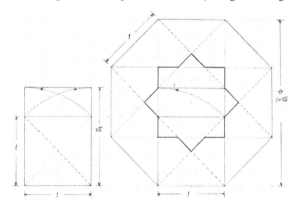

Fig. 5. √2 and octagon construction, graphite on paper

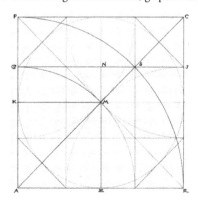

Fig. 6. "Sacred cut" and √2 geometry in the square, graphite on paper

While working with Verrocchio, Leonardo would have been directly involved in assisting his master in casting the sphere that was to be placed on the cathedral's lantern, and it isn't difficult to imagine Leonardo in the dome while this event was taking place, contemplating it all within the eight sails around him. Brunelleschi had worked with octagonal geometry to the point of perfecting it, and in making the octagon a permanent fixture in ecclesiastical building. He had secured the square based geometry of central system churches like the Old Sacristy, and with Santa Maria del Fiore, he established the permanency of the octagon. If we look at the prevailing geometry of this period, Leonardo tended toward shapes that related to and followed this trend. It is reasonable to assume then that the octagonal geometry of one of the great works of architecture and design stayed with Leonardo as he began his own investigations of central-plan architecture and octagonal structures.

At the same time, it should be remembered that Leonardo did not live in a vacuum, and these ideas and uses of octagons did not spring up fully formed in Leonardo's fertile imagination. In his review in *The Nation* (3/7/03) of the 2003 show entitled "Leonardo da Vinci, Master Draftsman" at New York's Metropolitan Museum, Arthur Danto points out that Leonardo was aware of the work of his medieval predecessors, which included artists, philosophers, builders, mathematicians, and religious writers, as well as those involved in the natural sciences and engineering. Danto cites the works of John Buridan, Albert of Saxony, and Nicholas Oresme as being among those to whom Leonardo was indebted. In historical terms then, Leonardo was not that far removed from medieval intellectual and practical life. It would be fair to think that he was also aware of the religious writings and thinking of St. Augustine and St. Bernard,[4] both of whom wrote at some length about the musical octave, relating to the number eight, and the double square, related to two squares, as is the octagon. (As Leonardo was also a musical enthusiast, the link seems fairly drawn.) Their writings were widely circulated in France and Italy immediately prior to Leonardo's birth, and were widely circulated within the circles he moved in during his own lifetime. The square, too, was held in great regard by philosophers, theologians, and intellectuals. Many thinkers of the twelfth and thirteenth centuries regarded it as the geometrical representation of the Godhead.[5] As we move then into the early sixteenth century, we find that octagonal geometry also occupied a fair share of the pavement designs for the Laurentian Library [Kappraff 2003, 213-235], indicating that it had held its own throughout the development of the Early Renaissance. So we see that square- and octagon-based geometry made a formidable and pleasing frame for Leonardo's creative life, and it appears that he welcomed the situation enthusiastically.

As mentioned, Leonardo's career ran roughly parallel to the development of the theme of the centrally-designed church. The Latin-cross cathedrals of the Middle Ages had made way for a renewed interest in the ancient Roman temple, a round, or possibly square or polygonal type of space. Specifically, the area of focus in the cross-type churches is where the two arms of the cross intersect. Many cross-type churches had incorporated the circle, square, and octagon as the shapes most successful in joining these intersecting arms (fig. 7).

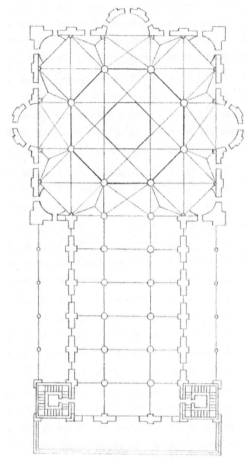

Fig. 7. Leonardo: Drawing for a central plan church MS. B, fol. 24 r, pen and ink

This focal point had developed into an area of exploration in and of itself. It can also be seen that these shapes were a natural result and the best solution to the geometric situation created by the intersection of the two arms. The next step in the evolution of sacred spaces in Europe would be the concept of a central plan, and the octagon, for all intents and purposes, was one perfect solution to the exploration.

Of the polygons, perhaps Leonardo recognized that the octagon, the decagon, and the dodecagon are those that appear to face in "all" directions and also have a relationship of four face tangency[6] with the square. If the octagon is selected for a sacred space, and as that sacred space is traditionally oriented to the Cardinal Points, four additional directions (Vitruvius's Eight Winds) could also be utilized. This would give the illusion of a welcoming facade gaining one entry and welcome to the space, regardless of one's point of origin. One of the concepts of the centrally-planned church was this welcoming, communal type of plan, and the octagonal structures that fulfilled this goal were beginning to be studied and developed now in Leonardo's Italy. In *Formal Design in Renaissance Architecture*, Michele Furnari writes:

The temple, known only through the many sketches of the Ashburnham Codex (fig. 8), belongs to a series of studies Leonardo made on the central-plan typology. As Heydenreich points out, this example is but one of many variations on the theme, ranging from simple plans based on a square or circle to richer, more complex ones based on polygonal shapes. These sketches bear precious witness to continuous experimentation with the concept of the central-plan system…

As Pedretti reports, "…As a result of his work on the tiburio of the (Milan) cathedral, Leonardo turned his attention to analyze the structure of a sacred building with a central plan, a typology in which the geometric center coincides with the central balance point of mass…" Leonardo began to combine different geometries and multiple models, starting from the core concept of a building composed of a pivotal inner space to which side spaces are added radially and symmetrically… He composed variations of ever increasing complexity: from square to circular plans, from polygonal to lobed ones. The central-plan concept was thus progressively enriched by the addition of new volumes to the central space… The intersection between the basic circular shape of the dome and other figures (squares, octagons, and so on) give rise to a variety of surfaces of cylindrical and spherical derivation [Furnari 1995, 36-7].

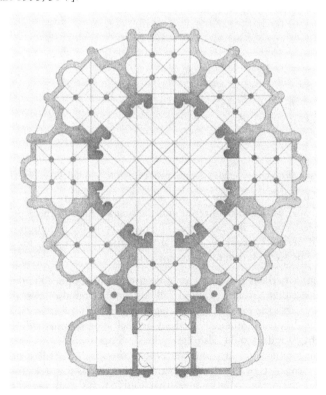

Fig. 8. Centrally planned church by Leonardo, MS. Ashb. 2037, 5-v; pen, ink, and graphite on paper

Fig. 9. Copy of Leonardo's sketches of cupolas (Sheet of eight drawings for raising a dome over a square base), MS. B, fol. 10 v, silverpoint on blue prepared paper

Leonardo was aware of all these aspects of eight-sided geometry, and he was also becoming aware of octagonal architecture in the plans, domes, cupolas (fig. 9), and decorative elements found in the architecture of northern Italy as well as in those cities with ports of entry from Eastern cultures. Venice was a major example, and we know that Leonardo made a point of traveling there. He would have known that Venetian builders were influenced in part by the arts and architecture of Persia, India, and the Middle East (fig. 10).

Fig. 10. Constructions of the star octagram and the octagon in the circle, graphite

These influences had made their way from the Holy Land and Asia to the Lion City through scholars, crusaders, and merchants. The variety of octagonal designs and their interplay were a key element in the arts, crafts, and architecture of these cultures, and the octagon held a valuable place, not only in Christendom, but also in Middle Eastern and Asian philosophy, spirituality, and symbolism. There can be no doubt that Leonardo was aware of all these things.

Octagonal architecture could also be found throughout other areas of Italy. Rome had a great share of samples, and Leonardo, while in Rome doing work for the Papacy, visited both Roman ruins and still extant Roman buildings (fig. 11). There, he had the opportunity to draw and study some of them. Of great interest to me were Prof. Pedretti's comments on Leonardo's book of drawings of old Roman architecture and temples [Pedretti 1985, 232-236], which indicate that some of these drawings influenced Leonardo's octagonal designs for the royal palace of Romorantin.

Fig. 11. Drawing of a Roman temple, Cod. Ambr., fol. 57, graphite on blue paper

As he began to circulate in the highest intellectual and politically powerful circles of his time, Leonardo would also have learned of still other famous examples of octagonal architecture. San Vitale in Ravenna was still high on the list of famous octagons in his homeland, as were other eight-sided masterworks from elsewhere in Europe, such as Aachen Cathedral and the Carolingian palace chapel in Germany, and St. Selpulchre in Cambridge, England. Leonardo also saw some of the octagonal architecture in other cities and towns in Italy during his travels. He was surely interested in the old octagonal "tower of Boethius" and the octagonal plan of S. Maria alla Pertica in Pavia (fig. 12) and he would have either seen a drawing of the structure by Giuliano da Sangallo while both architects were in Milan, or he may have actually discussed it with him.

Fig. 12. Construction of the plan for the Cathedral of Pavia, after Leonardo, MS B
fol. 55 r, pen and ink

It was also during this period that Leonardo took up the idea of designing portable pavilions for the Duke of Milan so that guests could be entertained during festivities outside the pressures of the city, or to be placed in the Duke's private garden within the formidable walls of the castello [Pedretti 1985, 63-73]. Giuliano had also done some work in this area, and both artists had seemed to find that the octagon and the square were the best geometric structures for this purpose. Of interest in this current study is the drawing Leonardo made of a plan and elevation of a pavilion in the garden of the Castello in Milan: *Cod. Atl.* fol. 3 v-b (fig. 13). There are no less than three octagons drawn out related to this project, indicating that Leonardo fully intended to have his octagonal studies reach fruition as an architectural reality for the Duke.

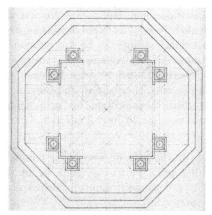

Fig. 13. Temporary architecture for festivals, *Cod. Atl.* Folio 3 v-b

Leonardo, as far as is known, had no real formal instruction in geometry. He did, however, have a small collection of books on the subject. Among these texts were Nicholas of Cusa's (1401-1464) *De transformationibus geometricus* and Euclid's *Elements*. He probably read some of the writings by the great Greeks: Archimedes, Theon of Smyrna, Apollonius of Perga, and perhaps even Pappas, Ptolemy, and Nicomachus. But texts like these would probably not have been of much help to him in his investigations specifically into the octagon, except where it may have been of some insight into approximations for π and for possible solutions to the problem of quadrature. (In early approximations for π, doubling the square – of which the octagon is the first step from square in the series – within a circle was a technique that was also examined as possibly an aid in squaring the circle.) However, treatises such as Leon Battista Alberti's *De re aedificatoria*,[7] and Francesco di Giorgio Martini's *Trattato do Architettura Civile e Militare*, which Leonardo annotated, were far more useful because artists and architects at that time were versed in geometry, not necessarily as a mathematical endeavor, but as a tool to be used in design, composition, perspective, engineering, and architecture. It would be from this position that Leonardo would most likely pursue his interests in things geometric.[8] (Leonardo's keen eye may have also picked up on one of Francesco di Giorgio's pages of geometry and writing, a page of regular polygons with the glaring omission of the octagon while drawing all of the other shapes; curious to say the least!) Those artists who also were gifted in the mechanical sciences might study geometry for other, more technical reasons, but in the context of the many drawings we find in Leonardo's journals, it is from an artist's perspective that we can best understand Leonardo's work with the octagon, for Leonardo was first and foremost an artist. By the time he worked with Pacioli, he had rendered countless octagonal constructions, and almost all of them freehand, another point that lends itself to an artistic appreciation of his geometric studies. Certainly, Fra Luca enlightened his mind with even more things numerical and geometric, but Leonardo learned them, heard them and saw them, foremost, through the heart and soul of an artist.

Regarding Alberti, when Leonardo examined his writings, he would have reviewed the drawings of the various orders of columns, the plan views of the orders being based on octagonal symmetry with alternating 45° radials emanating from the central vertical axis of the shafts. Of greater importance, I believe, would have been Leonardo's reading of Alberti's concise but nonetheless thorough explanations for several methods of constructing octagons in book VII, chapter iv of *The Ten Books*, as well as Alberti's efforts in determining the various rotational symmetries of the octagon. (However, it should be remembered that it was Leonardo, not Alberti, who actually put a theorem to these symmetrical operations.[9]) Alberti's influences may be read into some of Leonardo's drawings; the compositions, layouts, alternations, and ordering of the tribunals, chapels and apses would have benefited Leonardo in his own design quests. Among these influences, we find the following quote by Alberti that Leonardo probably read and considered utilizing:

> I would have nothing on the walls or floor of the temple that did not have some quality of Philosophy... I strongly approve of the patterning the pavement with musical and geometric lines and shapes so that the mind may receive stimuli from every side.

And what better representation of the musical octave, as mentioned, than the octagon, and what better way of using the octagon and square (the other possibility being hexagonal[10]) than by way of grids, patterns, and tessellations? Although Leonardo saw the idea of tessellation as a decorative device (figs. 14, 15), he also saw it as a solution to inhabitable space rather than religious, from palazzo to temporary pavilion (fig. 16).

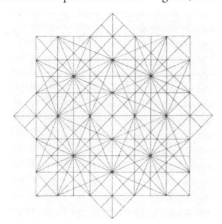

Fig. 14. Tessellation after Leonardo, *Cod. Atl.*, fol. 342 v-b, graphite and chalk on gray tinted paper

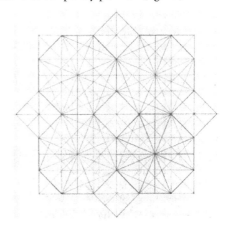

Fig. 15. Decorative patterns/geometric construction after Leonardo, pen and ink

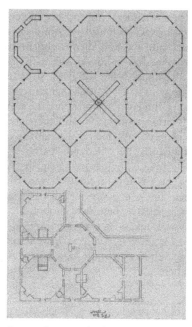

Fig. 16. Geometric study after Leonardo on multiple palazzo design, *Cod. Atl.*, fol. 349 v-k and v-c, silverpoint and sepia ink on brown prepared paper

Leonardo was long a lover of nature, a stalwart student of its structures and a tireless witness to its workings. With his great skills of observation, he would know early on that octagons, as a rule, and things referenced by the number eight, are somewhat rare in the natural world. Spiders, sponges, and octopuses aside, octagonal structures are mainly the logical result of the geometric relationships that exist and develop between: a) the square and the circle; b) the side of a square and its diagonal; c) combinations of these things along with the interplay of the diagonals and the vertical and horizontal midlines of the square (fig. 17).

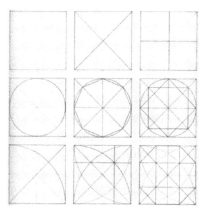

Fig. 17. Geometric constructions of square, √2 rectangles, and octagons, graphite

These things are geometric, not natural, and so with the few aforementioned items and with the major exception of the octave in music, we could say then that octagonal systems are mainly a human construct, and we find them frequently used in those contexts. In Leonardo's day, horoscopes were drawn not as they are today, but in squares, with diagonals and squares rotated through 45°, the basic structure of octagonal symmetry. European philosophers emerging out of the Middle Ages discussed the four seasons and the "Four Cross Quarters" marking the eight major points of a year, the four elements plus their four states diagrammed within an octagon, and the eight major winds of Vitruvius. On the subject of Vitruvius, we need only look at Leonardo's now-famous Vitruvian Man to grasp the idea. Vitruvius stated that the outstretched arms of a man are very nearly equal in length to that figure's height; that is, he fits into a square. And when his arms and legs are outstretched to the sides, he fits into a circle. If we reduce these two positions to mathematical signs, the human figure assumes the plus sign and the multiplication sign. Translated to the anatomy of a square, these two symbols are then the midlines and the diagonals in the square's make up. A contemporary of Leonardo's, Agrippa von Nettesheim (1486-1535), in *De occuta philosophia*, went so far as to keep the figures separate, and the two signs are even clearer in his drawings, as Agrippa also drew them with the geometry being discussed here.

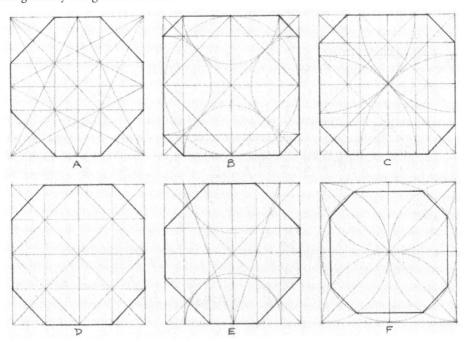

Fig. 18. Sheet of constructions of irregular octagons, graphite

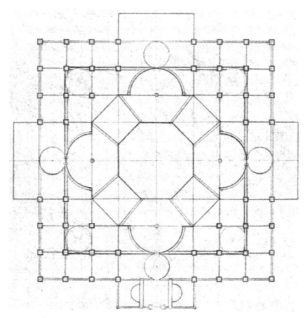

Fig.19. Drawing for a central plan church, *Cod. Ashb.* II, fol. 4 r; graphite on paper

It was a natural development for Leonardo to explore this octagonal geometry formed from the square. This is especially true if we are to believe Leonardo's own opinions about his rudimentary skills with geometric systems, that is, that he saw his geometric drawings as mentioned, as "geometrical recreations" [Cisotti, n.d.]. The square is universally known, and even people who loosely draw freehand will have, at one time or another, also sketched this very same geometric system of diagonals and midlines in the square. But for Leonardo, it was to be much more than what we think of doodling. Even if he halfheartedly believed his comment, the developed thumbnails were far more advanced than simply light sketching. A close examination of Leonardo's octagonal constructions shows that he was quite aware of the very different results obtained when working with the regular octagon as contrasted with irregular ones (fig. 18). His variations included eight sided figures using thirds of a square as well as using different rectangles to form crosses, connecting their corners to make the eight sides. As can be seen in the drawing in the *Codex Ashburnham II*, fol. 4-r (fig. 19), these irregular octagons are deliberate studies as opposed to random explorations, demonstrated convincingly in the beauty of the resulting structures.

Recently, I saw a documentary film on Leonardo's famous portrait of Ginevra de' Benci. In it, Professor Martin Kemp was being interviewed regarding his views on Leonardo and the portrait. Professor Kemp was quick to point out the similarities between the appearance of the spirals of Ginevra's hair and Leonardo's renderings of flowing water. He stated that Leonardo was always interested in finding connections between things seen in nature, and, in fact, he had a profound desire to unlock the mysteries of nature through the vehicle of visual art. Could this rarity of octagons in nature be the link between Leonardo's explorations of octagonal symmetry and the fact that there are so few examples of the octagon in the natural world? Simply, there is something rather unique about octagons, and Leonardo must have noticed it. So much so in fact, that even after he was exiled from Italy, in his last three years in France, Leonardo was still developing work on octagonal themes for the King of France while developing drawings for a royal palace at Romorantin. The studies are striking in their beauty and variety, and clearly demonstrate Leonardo's complete control and mastery over the octagon. They also support the fact that Leonardo, even at the end of his career, was still showing his preference for this shape, and the fact that the drawings are not simply reworkings of older studies adds yet further support.

IV Commentary on the Drawings

I have spent a good deal of time copying a number of Leonardo's centrally planned cathedrals, many of them octagonal in shape. With these, I also drew other octagonal geometry that he had rendered, unsure if it were a dome, a floor tile, an apse, a baptismal font, or a window, as he did not always label the drawings' utilitarian functions. On rare occasions, Leonardo took the time to ink up the compass nib and scribe a true circle or arc, leaving no doubt as to the geometric intention. More rarely, and quite helpful, were the countable, evenly spaced calibrations that could be followed, almost like graph paper. Sometimes, there were steps in alphabetical or numerical order. In some of the drawings, Leonardo's rapidity was such a key quality that I had no hope of unraveling the sketch without making assumptions and studies of my own. There were times I gave the construction a more finished look, with the intention of clarification rather than decoration or embellishment. Some sketches had precedents that I could refer to; some did not. Others were rooted in real buildings and objects, like the studies of the tiburio of the Milan Cathedral, or a tessellation found in Venice. There is a formidable array of recognizable, readable, and solvable octagonal geometry, and I have tried to pick the best of them, and, in my estimation, the more intriguing and the more beautiful of them, to present.

After reviewing about one hundred and twenty drawings, I entered about thirty or so of them into a working journal. I wanted to categorize the drawings and then do precise geometric drawings from the sketches in order to see them as formal architectural renderings or geometric constructions. This would be my best way to establish Leonardo's intentions and goals with his drawings. As a result of Leonardo's focus on centrally-designed plans, that is, with a clearly defined center and symmetry, some of the drawings look like Tibetan and Indian mandalas, others look like contemporary fractal geometry, some appear as tessellations, while yet others appear to be illustrations for mathematical principles. Of course, each drawing has at least a couple of these qualities present.

Because of the levels of complexity in the drawings, grouping them became necessary for me, and I came up with the following general categories:

1.   Mandalas and Tondi[11] (fig. 20a)
2.   Fractals (fig. 20b)
3.   Satellites: half-circles, circles, squares, octagons, etc. (fig. 20c)
4.   Alternations (opposing satellites) (figs. 20d, e)
5.   Irregular octagons
6.   Tessellations
7.   Domes and cupolas
8.   Quatrefoils and lunes (fig. 20f)

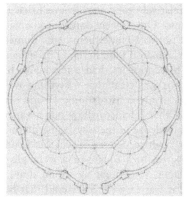

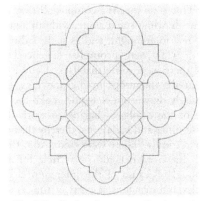

Fig. 20a. Plan for a many domed church (MS. B., fol. 35r); Venetian red pencil on cream colored paper

Fig. 20b. Project for a centrally planned church (Louvre 2282, reverse); brown ink on blue paper

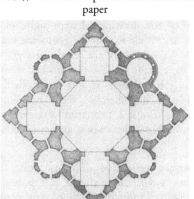

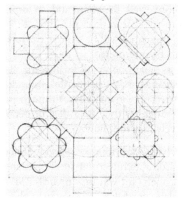

Fig. 20c. Plan of a many-domed basilica on a square plan with four apses (MS. B, f. 22 r); graphite, pen/black ink on paper

Fig. 20d. Geometric construction after Leonardo, graphite

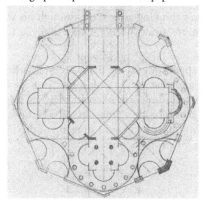

Fig. 20e. Centrally planned church. (Louvre 2282, obverse); mixed media on paper

Fig. 20f. Geometric construction after Leonardo (lunes) conte crayon

This is no means a complete listing, but it is a relatively good start from all that I saw. Some drawings stand alone, without a group to fit into. Others were done quickly, without any real forethought, and as such, I thought it best not to include them here. Sometimes his drawings were incomplete (a general criticism of Leonardo found even in his octagons!), and there were some that lacked any precision or direction. If I had a reasonable hunch about the intentions in the drawings, even though it was clear that everything was not precisely drawn out, I would make the adjustments in my constructions. To my benefit, one of the results of having to work this way was that I had to draw out a variety of possibilities and variations, thereby deepening my knowledge of the way Leonardo thought geometrically. There were times, too, that it was difficult to determine whether Leonardo had drawn an octagon or a circle for a baptismal area, or a square or a rectangle for a tribune. Because of the interconnectedness of shapes, sometimes he would draw both circles and octagons as parts of the same plan, and it was easy to understand that he was trying out different ideas with different shapes within one general framework. Because a straightedge was not always employed, parts of the plans did not align, and it was sometimes difficult to surmise where a wall was intended to end, a doorway to begin, what was connected to what, and so on. In a couple of the drawings, I kept some of the preliminary linework in the grid and construction lines visible so that the geometric system being presented will be more clear to the viewer. As a result, there are certain passages that had to be interpreted. In those cases, all attempts were made to have the finished drawing look as much as possible like the sketch on which it is based. It was not difficult to understand what Leonardo intended, even though his sketches were occasionally "incorrect", geometrically. Although perhaps appearing complex, Leonardo's drawings were not wrought with a great deal of detail nor an abundance of line work, and as such, were relatively easy to follow.

I have not found a good collection of finished geometric drawings done from Leonardo's architectural studies in any one location, and I believe that it is important to begin one. This was my intention in compiling drawings, some of which appear here (the complete portfolios of drawings that I did were donated to the Biblioteca Comunale Leonardiana in Vinci, where they can now be consulted). I believe that my experiences drawing geometrically can help others to gain some insights not only into Leonardo's creative oeuvre, but also into an understanding of octagonal symmetry. This understanding may provide some assistance in following Leonardo's thinking as he went through his work in this area of study, and it may suggest that more study be done specifically on the geometric aspects of Leonardo's art.

Before concluding, and on a lighter note, it is interesting to me that as there is a direct link between the octagon and the Maltese Cross, and that such a link might provide fuel for those who are interested in conspiracy theories regarding certain secret societies. Much has been made over the past several years about Leonardo's links to certain mystical fraternities, the Knights of Malta chief among them. I have included a drawing of this cross as used in the official Knights of Malta website (fig. 21).

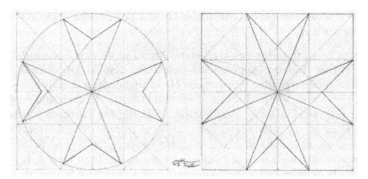

Fig. 21. Geometric constructions of a Maltese Cross, graphite

The second of the two is a variation. Indeed, the cross displayed on the site is constructed from the octagon, precisely. I say precisely because there are a variety of ways to construct what we call Maltese crosses, yet the official cross on the site is based on the regular octagon, and not on the more popular versions using the equilateral triangle/√3 geometry or the golden section.

V Conclusions

Leonardo da Vinci was a singular man of immense intelligence and ability. Even when selecting some detail of the master's work to examine, his enormity is still felt. It was this aspect of him that stayed with me as I looked over all the geometric sketches and constructions of octagons that he produced. My initial question regarding why he did so much with octagons above and beyond any other geometric shapes can perhaps be answered by historical precedents, the then-current trends, and the evidence of his sensitivities to symmetry operations. But I believe it was more than that. Leonardo's nature, abounding with curiosity and inventiveness, also played a major role. He recognized in the octagon its unique and particular structure in unifying circle and square, square to square, and square to rectangle. He recognized its relationships with other geometric shapes and systems, its potential for a wide variety of design possibilities, and the specific solutions it offered to spatial problems. Brunelleschi's dome demonstrated for him that among the geometric shapes at his disposal there were few shapes grander and stronger than the octagon. If Filippo's Dome had not convinced him, then Bramante's *opinio* [Pedretti 1985, 38-48] regarding the work Leonardo was doing on the Milan Cathedral solidified the octagon's strength and architectural qualities for him.[12] With all this, the octagon also satisfied the soul on symbolic and spiritual levels. This combination of factors gave Leonardo all he needed. There was really little else to do than to explore the myriad of possibilities present in octagonal geometry, and that is exactly what he did.

What we see in the journals is Leonardo at his best, the restless, curious, creative genius, making clay of the octagon, shaping it, bending it, rearranging it, and building with it. His approach to embroidery and knots had been similar, only with them he had recognized another quality of symmetry – that of tessellation – and chose the hexagon and the equilateral triangle for his system. In a similar way, he chose the octagon for his architecture. There was never any doubt in his mind that the octagon was not the best shape for him to work with architecturally. We do not see explorations with a variety of shapes. It was a prime example of Leonardo's "variations on a theme," and in this case, that theme was the octagon.

Speaking as an artist and geometer, it seems logical to me that Leonardo would select the octagon as that shape to examine most closely and develop, especially in light of his goals and philosophies regarding architecture. The octagon relates directly to his idea of the ideal central plan temple/church because of its relationship to the circle and the square, and how well the three work together. Additionally, Bramante and Leonardo had both spoken of the supremacy of the square, and the octagon is directly related to it, not only by containing two squares in its structure, but also because it is related to ad quadratum geometry and the 90° angle. Circular and square spaces had been built with regularity in antiquity, and the octagon had begun its presence only a short time before Brunelleschi and Leonardo took up the cause. Leonardo apparently did not want to get too involved with ad triangulum geometry, architecture that used equilateral triangles and hexagons. Had he, we would see more hexagonal temple designs in his drawings. By its nature, ad triangulum work presents a different set of building procedures and problems, procedures that Leonardo perhaps felt were too bothersome to deal with. The dodecagon, related to both the square and the triangle, was perhaps a bit too overwhelming and too similar to the circle. (In other words, why work with the twelve-sided figure instead of the circle?) The twelve-sided figure seems not to have been considered very much, perhaps because it is difficult to draw and handle, and too complex in its internal workings. It is possible that Leonardo saw the dodecagon as too grandiose and impractical, a shape without much architectural history. The pentagon and the decagon were problematic in that they are in conflict with the square and the 90° angle, and would have created numerous problems not only in rapid drawing and study, but also in designing and building. Simply stated, the octagon is relatively easy to draw, and once this aspect of it is mastered, it provides perhaps the greatest wealth of geometric possibilities and designs of all the regular polygons after the square. The octagon had a proven historical record, and it is also possible that Leonardo was simply carrying on the tradition of working with it, although with the great number of drawings he did with it, and the great number of variations he produced, I do think he realized that he was taking the shape farther than his predecessors did.

Even if we are to believe that Leonardo was being totally sincere about his lack of ability in geometry, the question still needs to be asked: Which aspect of geometry was he speaking about? The proofs, theorems, postulates and mathematical aspects (remembering that mathematics was not a forte of his either) of the subject, or those qualities that attract artists, designers, and architects? My guess is the latter. There is something about geometry and artistic creativity that go hand-in-hand. I believe that it was this aspect of geometry that attracted Leonardo more than any other. Having said that, I believe that the symmetry operations to be found in geometry ran a very close second in Leonardo's mind, and the symmetry operations of the octagon suited him best of all the simple regular polygons. This can be witnessed in a review of the voluminous geometrical drawings he did. His investigation into the octagon, as found in the drawing Windsor RL, 12700 (fig. 22) is analytical and scientific, to be sure, yet not really mathematical. (This same thing is true when we look at any of his sheets of geometrical drawings: there is little of mathematical notation and a great deal of drawing and quick sketching.) His analysis is one born of a natural curiosity, and one that lends itself to the use and development of geometric constructions in one's artistic and architectural work.

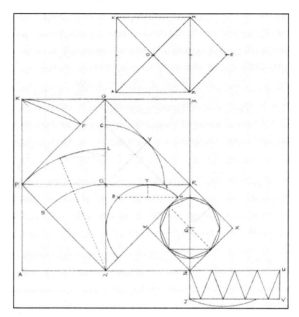

Fig. 22. Geometric construction after Leonardo, pen & ink

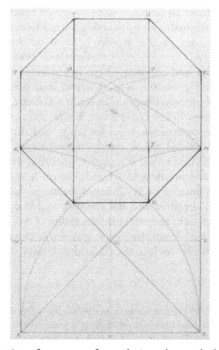

Fig. 23. Geometric construction of an octagon from ad triangulum and ad quadratum, by the author, graphite

Architecture stands between the twin pillars of art and number, and the two do not necessarily have to be separated to be viewed. They are linked by geometry, the visual presentation of numbers (fig. 23). However, when we look at Leonardo's codices, it would be safe to say that art, rather than mathematics, dominates his work. Whether Leonardo trusted the structural strength and integrity of the octagon as an architectural device because others' efforts had already demonstrated these qualities to him, or if he had known by mathematical study and discourse that the reliability of the octagon was above question, we may not ever know with certainty. We do not find a great deal in the way of this type of mathematical inquiry in his journals: that is, the use of Euclidean polygons as architectural elements based on their geometrically mathematical qualities. We do find a great many octagonal sketches and studies and sheets of other geometric systems and constructions, but without Euclidean notations and proofs. What we see is that Leonardo was comfortable enough with his knowledge of geometry in general and the symmetry of the octagon in particular to draw it freehand a great majority of the time. It was this overriding interest in the drawings that demonstrates for us his desire to see how the geometry worked. As mentioned, if we are to believe Leonardo's self-evaluation regarding things geometric, he arrived at his theorem on symmetry by drawing out the octagon in an almost endless variety of ways, and then observing those relationships. It was not calculation but Leonardo's observational skills and artistic ability that brought him expertise in working with geometric elements. This study may only reinforce the obvious suggestion that architects, artists, scientists, and mathematicians would do well to learn from Leonardo. He showed us that geometry provides us with a tool for artistic creativity and a sourcebook for design as well as securing for us the structural and mathematical aspects of masterful architectural forms. Still, there can be times when we would benefit from a study of a work of design from Leonardo's journals, not by studying the mathematics it contains, but rather by the geometry that was used. If that geometry, based on the user's expertise (or lack thereof) in mathematics, does not meet exact mathematical standards, it should reflect no less on its user's knowledge and abilities. There is a difference between geometry as a branch of mathematics and geometry as it is drawn and used in the physical world, especially in the practice of the arts. I believe Leonardo was aware of this. It is interesting to look at his approach to design in light of the fact that he did not believe himself gifted in the discipline of geometric constructions, yet he recognized that by its nature, geometry works wonders with a creative and artistic mind, and so he used it routinely in that spirit. In our efforts to be so specialized today, it seems at times that we may be creating a chasm between the philosophies and practices of art and science. It could be that geometry was more or less abandoned in favor of more advanced realms of thinking and practice before yielding all its treasures to the artist and the scientist, and it could be beneficial to take Leonardo's exploratory approach, and with his foundation of working with basic geometry. Perhaps Leonardo's drawings demonstrate that geometry, especially the simple, regular polygons and systems favored by ancient and classical architects, can provide all of us with a bridge to span that chasm.

Notes

1.   Prof. Carlo Pedretti [1985] has done a masterful job of presenting a splendid array of examples of these octagonal drawings taken from a wide variety of sources. The book is also one of the best historical documents on Leonardo's architectural life I have found.

2.   A famous, and rare, exception to his octagonal symmetry was Leonardo's hexagonal drawing for a mausoleum with a central plan in the Musée du Louvre.

3. See [Hopper 2000], various pages, especially Dante, pp. 154 and 178.
4. Cistercian architecture is based on the double square/ad quadratum approach to building philosophy.
5. See [von Simpson 1962, 49]. The entire text is an excellent source for the medieval concept of Order.
6. Any regular polygon divisible by four will be tangent to the four sides of a square. The octagon takes up more length on the square's sides than any other polygon.
7. By far, the best modern commentaries and translation on Alberti's treatise is [Alberti 1998]. [Tavernor 1998] is not to be missed either.
8. I have been unsuccessful in finding out if Leonardo had a copy of Piero della Francesca's *De prospectiva pingendi*, although it is somewhat doubtful. Had this been the case, the benefits may have been immeasurable for Leonardo.
9. Perhaps an interesting cosmic balance to the della Francesca/Alberti issue.
10. The second major tessellation system that Leonardo investigated for its architectural potential was the 30°/60°/90° tiling, but we have to wait until Borromini to see a full fledged effort in developing triangular and hexagonal geometry in architecture in Europe.
11. For a more thorough discussion on tondi, see [Arnheim 1982].
12. It should be recognized that Bramante was one of the greatest architects of the time, and Leonardo respected his opinions for this reason.

References

STIERLIN, Henri, ed. n.d. *India*. Architecture of the World, vol. 7. Lausanne ny): Editions Office du Livre,

ARNHEIM, Rudolf. 1982. *The Power of the Center*. London: The University of California Press.

CIRLOT, J. E. 1962. *A Dictionary of Symbols*. New York: The Philosophical Library.

CISOTTI, Umberto. n.d. The Mathematics of Leonardo. In *Leonardo da Vinci*. New York: Reynal and Co.

COOPER, J.C. 1978. *An Illustrated Encyclopaedia of Traditional Symbols*. London: Thames & Hudson.

HOPPER, Vincent Foster. 2000. *Medieval Number Symbolism*, Mineola, NY: Dover Publications.

FURNARI, Michele. 1995. *Formal Design in Renaissance Architecture*. New York: Rizzoli International Publications.

KAPPRAFF, Jay. 2003. *Beyond Measure*. Singapore: World Scientific Publishing.

PEDRETTI, Carlo. 1985. *Leonardo Architect*. New York: Rizzoli.

ALBERTI, Leon Battista. 1998. On the Art of Building in Ten Books. Trans. Joseph Rykwert, Neil Leach and Robert Tavernor. Cambridge, MA: MIT Press.

TAVERNOR, Robert. 1998. *On Alberti and the Art of Building*. New Haven: Yale University Press.

VON SIMPSON, Otto. 1962. *The Gothic Cathedral*. Princeton: Princeton University Press.

WATTS, Carol Martin and Donald WATTS. A Roman Apartment Complex. *Scientific American* **255**, 6 (December 1986): 132-140.

WATTS, Carol Martin. 1996. The Square and the Roman House: Architecture and Decoration at Pompeii and Herculanum. Pp. 167-182 in *Nexus: Architecture and Mathematics*. Fucecchio (Florence): Edizioni dell'Erba.

WEYL, Hermann. 1952. *Symmetry*. Princeton: Princeton University Press.

About the author

Mark Reynolds is a visual artist who works in oils, drawing/mixed media, and printmaking. He received his Bachelor's and Master's Degrees in Art and Art Education from Towson University in Maryland. He was also awarded the Andelot Fellowship to do post-graduate work in drawing and printmaking at the University of Delaware.

Mr. Reynolds is also an educator who teaches geometry for art and design students, and sacred geometry and geometric analysis for graduate students at the Academy of Art University in San Francisco, California. He has also taught linear perspective, drawing, and printmaking at AAU. He was voted Outstanding Educator of the Year by the students in 1992.

Additionally, Reynolds is a geometer, and his specialties in this field include doing geometric analyses of architecture, paintings, and design. Some of his studies can be found in the Nexus Network Journal at: www.nexusjournal.com

Mark has been on the Editorial Board of the journal, and also wrote 11 columns for, "The Geometer's Angle," discussing some of his discoveries and musings regarding the art and science of geometry.

Mark also lectures on his work in geometric analysis at international conferences on architecture and mathematics. Two of the most notable were at the Nexus Conferences in Architecture and Mathematics in Ferarra, Italy, in 2000 ("A New Geometric Analysis of the Pazzi Chapel in Florence"), and in Mexico City ("A New Geometric Analysis of the Teotihuacan Complex") in 2004.

For more than twenty years, Mr. Reynolds has been at work on an extensive body of drawings, paintings, and prints that incorporate and explore the ancient science of sacred, or contemplative, geometry. He is widely exhibited, showing his work in group competitions and one person shows, especially in California. His work is in corporate, public, and private collections throughout the United States and Europe. In 2004, Mr. Reynolds' had a drawing selected for the collection of the Achenbach Foundation of Graphic Art in the California Legion of Honor, and had 43 drawings accepted into the permanent collection of the Leonardo da Vinci Museum and Library, the Biblioteca Communale Leonardiana, in Vinci, Italy.

Mr. Reynolds is also a member of the California Society of Printmakers, the Los Angeles Printmaking Society, and the Marin Arts Council. Examples of his art can be found online by going to: http://www.markareynolds.com and http://www.caprintmakers.org and searching the Gallery.

João Pedro Xavier

Faculdade de Arquitectura
da Universidade do Porto
Rua do Gólgota, 215
4150-755 Porto, PORTUGAL
jpx@arq.up.pt

Keywords: Leonardo da
Vinci, representational
techniques, bird's eye
perspective, cavalier
perspective, axonometry,
centrally-planned
architecture

Research

Leonardo's Representational Technique for Centrally-Planned Temples

Abstract. Leonardo invented a new technique of representation which combines the building plan and a bird's-eye perspective of the whole into a single system. Bird's eye perspective may have developed out of cavalier perspective, and instances pre-dating Leonardo can be found, but not used in the same way as he employed it. Though not pre-axonometric, Leonardo took advantage of axonometric representation's capacity to construct/deconstruct an object into its component parts in order to clarify fitting and functioning. This paper investigates the originality of the technique and special relationship with his research on centrally-planned churches, while examining it in the context of contemporary developments and architects.

It is hard to deny that Leonardo, among many other inventions, should also be credited as the inventor of a new technique of representation, especially adapted to his research on centrally-planned churches, which combines, as a system, the building plan and a bird's-eye perspective of the whole, as in MS 2307, fol. 5v. (fig. 1).

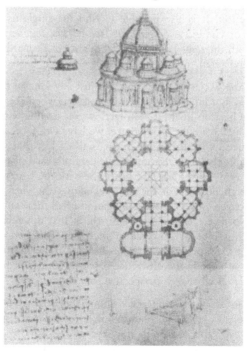

Fig. 1. Leonardo da Vinci, plan and bird's-eye view of a centrally-planned church (MS 2307, fol. 5v)

Nexus Network Journal 10 (2008) 77-100   NEXUS NETWORK JOURNAL – VOL. 10, NO. 1, 2008 77
1590-5896/08/010077-24 DOI 10.1007/ s00004-007-0057-7
© 2008 Kim Williams Books, Turin

This was already pointed out by Murray, who wrote:

> …in the great majority of the drawings in the architectural treatise Leonardo presents a plan and a bird's-eye view, so arranged that the maximum amount of information about a building, both externally and internally, is given in two drawings (…) [Murray 1978: 62].

Previously, Lotz had connected the simultaneous use of these two kinds of drawings with the representation of centrally planned temples [Lotz 1997: 94].

For both Murray and Lotz, the great innovation is the bird's eye-view.

Murray relates this type of perspective with Leonardo's anatomical drawings, asserting that the use of the same technique in architecture "opened a whole field of new possibilities" and completes this statement with this challenging commentary:

> … it is probable that Bramante inspired him to invent this new form of draughtsmanship in connection with architecture and in turn, Bramante's whole cast of thought would have influenced by the new technique of visualization [Murray 1978: 62].

Lotz maintains that the bird's-eye view contains the elements of a perspective construction that will be known later as "cavalier perspective", a representational method developed in the last quarter of 1400, according to him.

In spite of these respectable opinions, I believe Leonardo's genuine and productive innovation was actually the systematic combination of a bird's-eye view and a plan in architectural representation. It is also possible to find the single use of the first kind of drawing in Francesco di Giorgio's *Trattati di Architettura Ingeneria e Arte Militare* (c. 1485), although not with the same coherence and quality of Leonardo's sketches, who was undoubtedly a gifted draftsman.

We should specify that we are not dealing with just any kind of bird's eye-view, which could be simply defined as a kind of perspective produced when the object the observer is looking at is above the horizon line, HL.[1] In fact, we are dealing with central perspective, which means that if we take a cube as reference, we have two faces parallel to the picture plane that are, consequently, in true form. But it is also a fact that in these drawings this supposed cube is laterally placed in relation to the central visual plane, which gives us the possibility to see another of its vertical faces, this one with the correspondent homological transformation. The high position of the observer allows the view of the top face, also transformed by perspective. As we see, the conjugation of all these factors leads to the most usual form of cavalier perspective, and we might think that this relationship would grow proportionally as the distance of observation increases. But here is the problem: the existence of a point of view placed at a finite distance from another one placed at the infinite is crucial, as it is actually the difference between linear and parallel perspective, or in other words, the difference between perspective and axonometry.

Massimo Scolari, in an interesting essay about the origins of axonometry [1984], contextualizing its inherent qualities in philosophic terms, talks about a voyage that begun in "Plotinus' interior eye" and arrives to the "eye of the Sun", to use Pietro Accolti's language,[2] becoming a few years later the "point at infinity" in Desargues's projective geometry. In any case, to Scolari, the development of axonometry coexisted with that of

---

perspective. This means we had a combination, from the first attempts to represent spatial depth in a plane until the sedimentation of both representational systems in scientific terms, begun in the fifteenth century, of two ways to view or represent the world, or two ways of facing or thinking it: this interior eye vision found in axonometry and the perspective "body's eye" a former *mimesis* instrument to reproduce nature. And, as Alan Colquhoun explains,

> [This] debate between Euclid's and Democritus's theories of vision, with their rival aesthetic implications was resolved, in the course of 16th and 17th centuries by dividing the field of representation with two discrete parts. Mimesis in general became absorbed by perspective, while parallel projection and axonometry were preferred wherever the criterion of value was descriptive accuracy [Colquhoun 1992: 17].

As far as I know, the first rigorous cavalier perspectives appeared in Piero Della Francesca's manuscript *Libellus de Quinque Corporibus Regularibus.* It is not surprising that Luca Pacioli also utilized this kind of perspective in his own drawings for *Divina Proportione.* An example could be Piero's drawing of an icosahedron inscribed in a cube, which Pacioli repeats in an engraving suitable for printing (figs. 2 and 3). Curiously, Leonardo's drawings for Pacioli's treatise are not cavalier but bird's-eye perspectives (fig. 4). I will return to this further on because I believe this is rather significant.

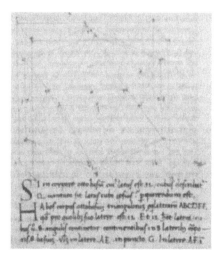

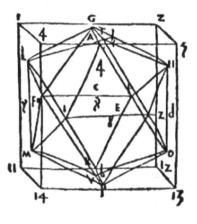

Fig. 2. Piero della Francesca, icosahedron in a cube. From *Libellus De Quinque Corporibus Regularibus,* fol. 40v

Fig. 3. Luca Pacioli, icosahedron in a cube, *Divina proportione,* Libellus, fol. 15r

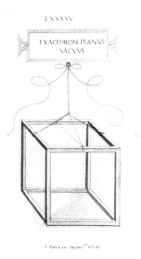

Fig. 4. Leonardo da Vinci, Hexahedron (cube), from Luca Pacioli's *Divina proporzione* (1509)

So, polyhedral representation or, more widely, the representation of spatial geometry will come to be preferred to cavalier perspective drawings. It seems, though, that when the objects of pure mathematics become objects of a scientific approach in the field of representation we tend more to axonometry than to perspective in spite of all the astonishing perspectives of polyhedrons presented in perspective treatises through the sixteenth and seventeenth centuries.

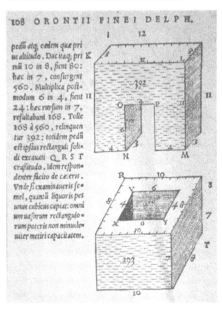

Fig. 5. From Oronce Finé, *Liber de Geometria Pratica*, 1544

But this, I believe, with a few exceptions, was more due to virtuosi and artistic purposes or pure pedagogical reasons, such as the use of such shapes that could facilitate the learning of perspective, than for a mathematical approach to geometry itself. However, it is interesting to note that Piero was responsible for the beginning of a tradition leading to our days, as cavalier perspective is still the main tool for visualization in mathematics literature. One of the reasons for this longevity could be its connection with an intuitive process of spatial representation, its simplicity and readiness of use and, of course, the possibility of obtaining true measurements. It was Oronce Finé (1544) who first scaled the solids edges legitimating the future designation (dated from the nineteenth century) of this representational system – axonometry (axis measurements) (fig. 5).

But the main reason, in this case, could be perhaps the consideration of the object as itself as axonometry gives us a key to "read" it without the subjectivity inherent in a personal eye. So, in axonometry, there are not eyes but a conventional one (actually, there is more than one), placed at infinity, which allows a close approach to the object, to its own characteristics and properties, without the veil that visual perception could impose in between.

These drawings, unequivocally rigorous and as thus perfect axonometric projections, achieved a long time before the theorization of this representational system (but this happened with all systems of representation), cohabit with the kind of bird's-eye views, used by Leonardo and Francesco di Giorgio, which are usually considered as pre-cavalier drawings, and thus pre-axonometric ones, as actually they are not far from geometric cavalier perspective. Of course, there is the lack of parallelism of the perpendicular lines to the picture plane, as they converge to the central vanishing point (the *punto centrico* as Alberti baptized it). There is also the impossibility of picking up true measures. Anyway, the considerable distance of the observer, placed at a point where birds usually are, in a middle stage between man's view in a standing position – that is, with feet on the floor – and the infinite distance of the eye of God, gives a complete three-dimensional view of the object and the possibility of describing it in a synthetic way, a quality typical of axonometry.

In the specific case of Leonardo, I think he deliberately proscribed parallel or cavalier perspective. He wanted to do exactly what he did: nothing other than bird's-eye perspectives!

What gives me permission to state this? Fundamentally, two things. First of all, Leonardo lived in the world of perspective, which he proclaimed several times as the unique instrument that makes it possible to get knowledge of nature and all the environment (including all the artifacts it is possibly to find in it or invent for it). Even in the field of perspective it is possible to verify that Leonardo never forgets that the point of view is actually the human eye, in contrast to Piero della Francesca's mathematical approach, where he considered it more a single geometric point (cf. [Cabezas 2002: 147]), although his preference for bird's-eye views could lie behind the image of the winged eye, shown in the verso of Matteo de Pasti's portrait medallion of Leon Battista Alberti, if we accept that this emblem, a symbol of perspective, inspired the kind of imaginative vision which came to encompass all his aesthetic ideas in its gaze, as noted by Gadol [1969: 69]. Secondly, as I remarked before, and I believe this is determinant, his drawings of polyhedra for Luca Pacioli were perspectives and not cavalier, as Luca Pacioli's own drawings were in the footsteps of Piero. So, he knew perfectly that kind of perspective, where lines he actually saw as convergent strangely remained parallel against the evidence of his senses. And this, for a man who had an absolute faith in experience, could never be tolerated!

Thus I believe that, with Leonardo, we are not dealing with a drawing production classifiable as pre-axonometric in the sense of a primitive stage of that kind of representational system. Such consideration is valid for the prior tradition as in Taccola's pre-cavalier perspectives of mechanisms (fig. 6), which both Francesco di Giorgio and Leonardo followed (fig. 7). But if in the engineering drawings made by Taccola and Francesco di Giorgio some detectable inconsistency can be attributed to their incapacity to control three-dimensional representation,[3] which does not compromise at all its complete efficacy for displaying their functional and constructive aspects, I think that with Leonardo things are completely different.

Fig. 6. Mariano di Jacopo (il Taccola), Sailing-car, *Liber Tertius de Ingenis ac Edifitiis non usitatis*, 1433

Fig. 7. Leonardo da Vinci, crane to lift heavy weights, *Codex Atlanticus* fols. 30v/8v-b

Once again, without leaving his beloved perspective in this form of bird's-eye views (a technique he mastered as no one ever before), he took advantage of another quality of axonometric representation: the capacity to construct/deconstruct an object into its component parts in order to clarify fitting and functioning, which was another main reason to keep axonometric drawings near to engineering production or architectural construction details.[4] If we look carefully at the bird's-eye perspectives he used to explore all the fields he was interested in, which were many, we can detect that the point of view he usually chose was not as high as it is now in conventional cavalier perspective.[5] As regards his architectural drawings, we verify no lack of realism, since the views of a single building or an entire city are obtained as if he were looking at them from the top of a hill, as in the first landscape we know of his (fig. 8).

Fig. 8. Leonardo da Vinci, landscape, 1473. Florence, Uffizi Gallery

Particularly, if we think of his centrally-planned churches crowned by a dome, we can easily perceive that he has drawn them as if he were viewing the dome of Santa Maria del Fiore from the neighboring campanile by Giotto, or the church of Santa Maria degli Angioli from the lantern of Santa Maria del Fiore, which were determinant referential models for his research on centrally-planned space. Carlo Pedretti goes even so far as to suggest that this view from a high point to three-quarters of the Tempio degli Angioli, similar to a view of an actual wooden model of the church placed on a table, must be related to Leonardo's use of this kind of perspective in architectural drawings [1981: 14].

Even in his drawings of cities, Leonardo remained loyal to bird's-eye perspective (fig. 9), while Francesco di Giorgio was developing for his fortifications a kind of drawing where the front view and the plan tend to appear, simultaneously, in true form (fig. 10). Faced with the need to display clearly the geometry of his plans for fortified cities, Francesco di

Giorgio intuitively discovered a drawing method which proved to be perfect for representing military architecture and, later on, architecture in general. That is why the rigorous form of this kind of perspective, today called "planometric projection", was first known as "military perspective", or *prospettiva soldatesca* as Girolamo Maggi and Jacomo Castriotto called first it. According to Scolari, parallel perspective as an alternative to Renaissance perspective was clearly presented, for the first time, in their work *Della fortificazione delle città* (Venice, 1564).[6]

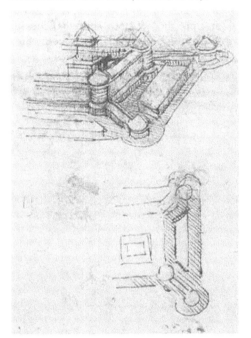
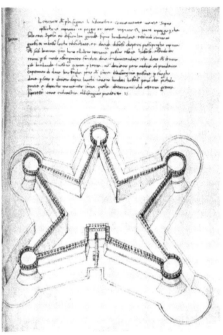

Fig. 9. Leonardo da Vinci, bird's-eye view of a fortress (c. 1504). Manuscript 8936, fol. 79r, Madrid, Biblioteca Nacional

Fig. 10. Francesco di Giorgio, Pentagonal star fortress, *Trattati di Architettura Ingeneria e Arte Militare*, c. 1485

Returning to the new technique brought into architectural representation by Leonardo, we should now try to understand its novelty and special relationship with his research on centrally-planned churches, while examining its contextualization.

In a way we may consider Leonardo's system – plan plus bird's eye perspective – as a personal synthesis of Vitruvian ideas on architectural representation, since for him it was enough to use only two kinds of *dispositio*, as Vitruvius says: the so-called *ichonographia* (ground plan) and *scenographia* (perspective). It seems evident that the missing element of the triad, the *ortographia* (elevation), could be dispensed with, as it was implicitly included in the frontal bird's-eye views he used. In any case, as far as the position of the observer is concerned, Vitruvius's *scenographia* seems to be a bit different. There is no doubt that he is referring to central perspective, certainly in its pre-stage (there were no means, at that time, to establish correctly the diminution of distances in depth), when he says, "…perspective is the method of sketching a front with the sides withdrawing into the background, the lines all meeting in the centre of a circle" [Vitruvius 1970: I, ii, 15]. But as he only refers to the possibility of seeing the sides of the building and not the roof, we can infer that the observer was not at the high level where Leonardo always preferred to place him in order to control three-dimensionality in a single drawing. And, as we know, he was particularly interested in the design of the roof, since the shape, size and disposition of the domes, apart from their constructional aspects, were questions of main importance in ecclesiastic architecture. Further, according to Richter, it seems that these church drawings were expressly designed for a "Tratatto delle Cupole" that he envisioned writing [Richter 1970: 38].

I have already presented an ensemble of possibilities derived from such bird's-eye positioning in general: its three-dimensional ability to give a synthetic overview of the whole, the focusing on the object in itself, contributing to the possibility of underlining its geometric profile and showing its parts and the way they are spatially assembled and combined in order to compose the whole. And, concerning the measurements, we should also add, at least, the possibility of getting proportional measurements in the frontal views, since its form remains unaltered.

On the other hand, a plan drawing (*ichonographia*),[7] in addition to its main role in functional aspects of accessibility, circulation and interconnection, which are related to men's horizontal movements through space, is like an engraving on the ground of the building's geometrical matrix – a perfect mirror for its mathematics. And, actually, it is also from the ground up that it will grow.

In combining a plan – sometimes a mere diagram or just a scheme – with the 3D possibilities of bird's-eye perspective, Leonardo found an efficient instrument for sculpting a building in space, a way to control its growth through its final stage, ensuring its geometrical integrity from the floor to the roof. With these two drawings he could also give the maximum amount of information about it, as Murray underlined. But it is easy to prove that this is especially true in the case of centrally-planned temples, and even there we shall have opportunity to point to some indeterminations and unsolved questions.

It is evident that what makes this information "almost enough" is actually the central symmetry of the building.

Bird's-eye perspective, simultaneously presenting the front elevation and a lateral one, is implicitly telling us what the conformation of the other two sides not shown is. If any doubt remains, it is sufficient to look at the plan to confirm this. The elevation in frontal view also indicates the proportional relationship between horizontal and vertical measures of the façade, which is extensible to them all.

One of the façades, often the one in frontal view, is subtly different from the others because it is where Leonardo locates the entrance. Even so, it is difficult to speak about the existence of a main façade, since Leonardo's concern for emphasizing centrally-planned space forces him to avoid distinguishing any kind of direction. The door is there, but only to connect exterior and interior. It is not by chance that he sometimes even omits it. Further, sometimes we don't even know where it could possibly be placed, as in the plan for an octagonal church with round chapels in each side (fig. 11, upper right). To contradict this, there is also a more elaborated octagonal plan with Greek-cross chapels on each side (see fig. 1), with an exceptionally developed narthex. But here, as it does not appear in the bird's-eye view, it seems that it was added only in plan, as a possible solution for solving the difficulties inherent in the previous plan scheme. So, the prevailing feeling is that the narthex is, in this case, more a matter of concession than conviction and perhaps a result of looking at ancient examples. I believe that all these reasons could explain Leonardo's reluctance to use Alberti's system, which prescribed the use of drawing an elevation (*ortographia*) together with the plan, to which Rafael later added the use of the section, since he did not want to give preference to any of the façades of his centrally-planned temples.

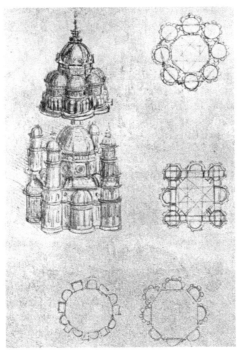

Fig. 11. Leonardo da Vinci, drawings of churches. Manuscript B, Institut de France, Paris

Due to central symmetry, the other thing a bird's-eye view makes evident is the whole configuration of the roof, where the main element is always a central dome. Its shape is generally octagonal or hemispherical. Sometimes there are smaller domes around, which can be intercalated by towers, and there can also be apses, sub-apses, niches and so on, with their corresponding domes. Whatever the building top is, even without seeing the whole, we are able to get all the information we need about it, as every element is symmetrically disposed in relation to a central vertical axis. This vertical axis, which runs from the center of the ground plan to the top of the lantern of the main dome, directs the connection between the ground and the roof plan and thus perspective displays three-dimensionally in the exterior the result of interior spatial shaping, particularly the balanced configuration of its different kinds of domes.

In some cases there is a lack of information about the interior space, which could only have been overcome with an orthographic section or, alternately, with a perspective of the interior.

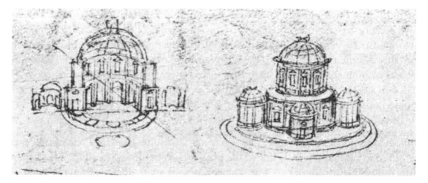

Fig. 12. Leonardo da Vinci, Circular temple with four circular chapels. *Codex Atlanticus,* fol 205 v

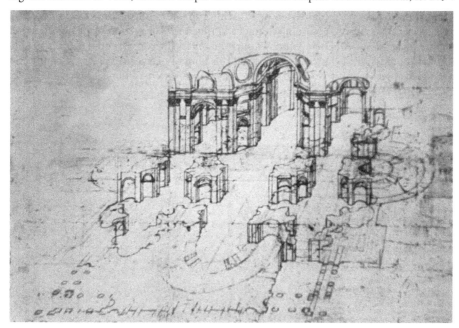

Fig. 13. Baldassarre Peruzzi, design for San Pietro di Roma. Firenze, U2A

While the resort to a section had been used since medieval times, the use of perspective in architectural representation, as pointed out by Lotz, made its appearance in Filarate's treatise, in the form of perspective-sections, and, in spite of Alberti's advice, the use of perspective, developed mainly in painting, became irresistible to architects as well. That is the case of Giuliano and Antonio da Sangallo, Francesco di Giorgio and Bramante.

Leonardo is an example of this, a very special one since he used perspective in a very peculiar way, as we have seen. But the fact is that he also used perspective sections and perspectives of the interior space. Sometimes the front views of his bird's-eye perspectives are actually sections, but there are also some situations where the observer is placed at a normal eye level, as if he were looking into the sectioned part of the building or into its interior.

One of the two drawings of a circular temple with four surrounding circular chapels is a bird's-eye perspective section (fig. 12). Here, it is undeniable that the section through one of its two symmetric vertical planes and the footprint of the half of the building not shown increases the information about it, even dispensing with the actual plan. This kind of strategy was announced Peruzzi's astonishing deconstructive bird's-eye perspective of his project for St. Peter's in Rome (fig. 13). But, evidently, axonometric drawing gradually appropriated this kind of approach for an architectural object, since it clearly departed from the field of naturalistic observation.

From the second kind of perspectives there are also some drawings regarding centrally-planned space in temples, even if the whole plan is classifiable as a basilica. All of them are related to the problem of covering a square space with apses on at least three sides, as the fourth one could be the basilica nave.

The drawing that has been related to an architectural idea for the cathedral of Pavia is a typical perspective section, according to tradition (fig. 14). The section plane, which acts as a picture plane, or an Albertian window, puts the observer aside the interior space in spite of the low point of view chosen. I believe that this kind of perspective is also in conformity with direct observation of wooden models, which were often sectioned along one of the vertical planes of symmetry, in order to show the interior.

Fig. 14. Leonardo da Vinci, perspective-section of a church apse. *Codex Atlanticus*, fol. 7v-b

The other perspectives are quite different, because the observer is now immersed in the space. That is the case of a drawing for a proposal of a possible renovation of San Sepolcro in Milan (fig. 15).[8] The same attitude is found in a series of interior perspectives regarding a project (or projects) for a Greek-cross plan church (fig. 16). Such kind of perspectives, which Leonardo uses to control the articulation of different kind of elements that define space, are also in correspondence to a possible direct experience of space by the observer.

Fig. 15. Leonardo da Vinci, studies for a church apse, c. 1488. Windsor, n. 12609v

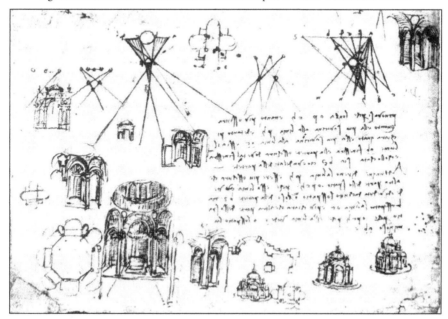

Fig. 16. Leonardo da Vinci, studies for centrally-planned churches, c. 1508. *Codex Atlanticus*, fol. 37v-a

I believe that this is the point where perspective really meets architecture, in the sense that we are allowed to state that space was shaped and thought of according to the way the spectator will see and feel it.

Elsewhere I have explored the close correspondence between the development of central perspective[9] and the research for centrally-planned space, in both urban and building planning, during the Renaissance [Xavier 1997]. The first known material evidence of such a tendency only came with the high Renaissance, as is the case of these perspectives of Leonardo's as well as the others attributed to Bramante, to which we should add the earlier panels of the *Ideal City* depicted by Urbino's circle (a presumed joint-venture between Luciano di Laurana, Francesco di Giorgio and Piero della Francesca) where Bramante, as a matter of fact, worked during his apprenticeship.

Returning to Leonardo's centrally-planned churches and his representational system, it can be seen that the information given is in fact sufficient in the majority of the situations, despite the exclusion of a section or a perspective of the interior.

First of all, we must take into account that we are working with sketches. So we have to deal with the hesitations, the superposition of different schemes or hypotheses, some imprecision in the drawing of lines or just their mere fading due to the passing of time. In any case, we should see that as a privilege, because it allows us to follow a mind at work through drawing, in this case seeking different and ingenious solutions and variations around the theme of centrally-planned space in temples. Leonardo himself proclaimed, *il disegno è una cosa mentale*, drawing is a mental thing, and in his case, this is absolutely true! So, we are not in the presence of drawings for execution nor was that their purpose. It appears that their main use was to be part of a theoretical treatise on architecture, which does not mean that these drawings could not reach a rigorous definition in order to produce a wooden model. Such models were still indispensable elements for preparing the project for construction.

Secondly, we have to consider that Leonardo was manipulating a stylistic grammar that was understandable to his fellows and followers. And, in fact, he uses a limited number of spatial cells, which were not unknown, but as he experiments with a wide range of possibilities and combinations, he arrives at some original results. However, it was not Leonardo but Bramante who was actually the first one who could amplify these achievements in executed buildings where special attention to proportion and the clarity of classical language became evident. This is also true for his designs, especially the one for San Pietro in Rome.

With the awareness of these limitations, it is possible to re-draw some of Leonardo's temples, which testify to the univocal sense of his representational system in such cases. There are even some models exhibited in museums, and it also possible to construct them virtually.

This is the case, for instance, of a temple with a regular octagonal plan surrounded by octagonal chapels on each side that are inscribed in a larger polygon with sixteen sides (fig. 17). There is no doubt that the main space is an octagonal prism crowned with an octagonal dome and we can be sure that the same occurs in the subsidiary chapels. There is also a similar plan where we find circular chapels on each side of the main octagon, but this time they are exterior, and show surrounding small apses that became niches in the interior side (see fig. 11, upper right). A final possibility is present in the very same plan, as there is a homothetic octagon whose sides cut these round chapels at the middle, transforming them into semi-circular apses. This is confirmed by a scheme present in the bottom of the same page. As we can see, the smaller apses remain. The corresponding bird's-eye perspectives confirm that these circular chapels are cylinders crowned with hemispherical domes and the apses half-cylinders covered with half-hemispheres, as shown more schematically on another page (fig. 18, above left).

Fig. 17. Leonardo da Vinci, octagonal church with eight octagonal chapels around (Model - Museo Nazionale della Scienza e della Tecnologia "Leonardo da Vinci"; drawings - MS B, fol. 21 v)

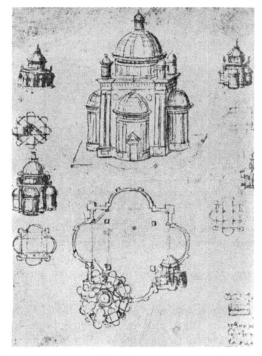

Fig. 18. Leonardo da Vinci, Drawing of churches MS BN. 2037, f. 3v. Paris, Institut de France

There is also another family of plans generated from a regular octagon to which correspond, once again, a prismatic space covered by an octagonal dome. The church presented in MS B, fol. 22 r is an example of this kind of plan. Here again there is no any possibility of misunderstanding (fig. 19). Considering the octagon as the result of the intersection of two concentric squares rotated 45º, we arrive to the whole plan with an *ad quadratum* growth of each square. Along the way, corresponding to the corners of each square, we will find four chapels and four apses. One of the exterior limiting squares, which cuts the apses at the middle, corresponds to the dominant quadrangular prismatic mass of the building from which the main dome and the four domes of the chapels are detached. In the example selected, the chapels are quadrangular prisms crowned with hemispherical domes perched on cylindrical drums; the half-detached side apses are cylinders with hemispherical domes.

Fig. 19. Leonardo da Vinci, plan generated from a central octagon with an *ad quadratum* structure MS B, fol. 22 r., Paris, Institut de France. Geometric overlay by the author

But the problems arrive with the squared plan, which is actually related to the Greek-cross plan, to which corresponds in most cases a hemispherical dome with a cylindrical drum, although it can also be octagonal.[10] It is not by chance that all the interior perspectives we have are in relation to this plan, but even so, not everything can be disclosed.

It is sufficient to look at the several plans presented in MS 2037, fol. 3v to find the first difficulties (see fig. 18). The larger plan to which the larger bird's-eye view above presumably corresponds does not in fact match. Comparing the dimensions of the dome to the visible half-domes of the apses, we conclude that it should cover the whole space, which is incompatible with the presence of the piers in the plan. There is a more schematic plan at left, and another, even more schematic one at the right of the larger bird's-eye view, in which the piers do not appear, but we cannot be sure if this is a consequence of the simplicity of the sketches. What is certain is that the plan that matches the bird's-eye perspective cannot have piers. But there cannot be any kind of connection in the corners either, as is also shown in the plan, suggesting the definition of a semi-regular octagon inscribed in the square. Anyhow, there is also marked in the same plan (upper left corner) a short arc of a circle that could correspond to part of the projection of the dome shown in perspective. If, on the contrary, we stay with the four piers, the dome necessarily has to be shorter and should be inscribed in the square defined by these piers. But if it is the semi-regular octagon that remains, then the dome can be maximized again and here we have the interesting question posed by the spatial transition between a semi-regular octagon and a circle, a problem presented and perfectly solved in Bramante's project for San Pietro in Rome.

Considering all these possibilities, and others not yet mentioned, we begin to believe that from that plan it is possible to extract different plans, or more exactly, different variations of the same type of building, which is a clear indication that Leonardo was aware of such implications and was trying to find solutions for the problems posed by each. So, here again we see his mind in action.

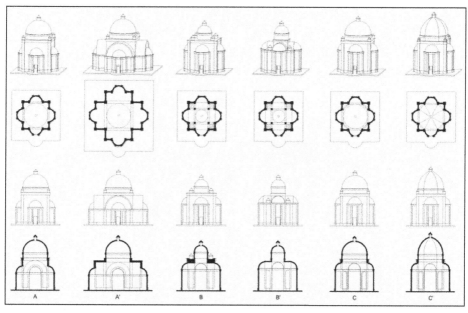

Table 1. Types of centrally-planned churches derived from MS BN.2037, fol 3v.
Paris, Institut de France

I have tried to reconstitute these possibilities diagrammatically, where each type (or sub-type) is presented in Leonardo's preferred system – plan plus bird's-eye perspective – to which I added a section and an elevation (Table 1). Because I could utilize CAD software to make these drawings, I also constructed virtual 3D models which we can (virtually) enter. With this instrument, the remaining doubts about the interior space become more evident.

All types take as their point of departure a squared plan. By subdividing each side into four units we get a net of sixteen squares, which underlies the dimensions of all the spatial elements present in all the types. It seemed to me that this grid fit, more or less, in the majority of the sketches, particularly the larger plan of MS 2307, fol. 3v (fig. 20).

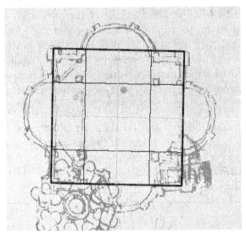

Fig. 20. Squared plan grid. Geometric overlay by the author

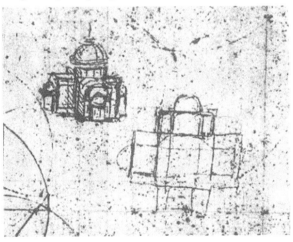

Fig. 21. Leonardo da Vinci, Greek-cross church. *Codex Atlanticus*, fol. 362r-b, v-b

Type A corresponds to a neat squared plan with one semicircular apse on each side. The main space is almost cubic and a hemispherical dome perched on a cylindrical drum covers it. The apses are cylindrical and crowned by half-hemispheres. In the exterior there are four pinnacles in the corners of the cubic prism, which are actually small *tempiettos*, as in the project for the Cathedral of Pavia. The transition between the square plan and the circular base of the cylindrical drum could be certainly solved with spherical triangles. The building that is most similar to this type of scheme is Bramante's Santa Maria delle Grazie in Milan. Both realities must be taken in consideration in order to explain Santa Maria della Consolazione in Todi. Here, however, the apses are wider, which brings the plan closer to a Greek cross.

Type A' is configured as a clear Greek-cross plan, but the exterior appearance of the vaults of the detached arms of the cross is not convincing (fig. 21). Perhaps it could be covered with a saddle-roof with a pediment at the end, as is shown in a perspective in the *Codex Atlanticus*, fol. 37 v-a (second from right at the bottom of the page) (see fig. 16). The most similar examples in actual buildings are Giuliano da Sangallo's Santa Maria delle Carceri in Prato (1485-95), although apses are absent there, and San Biagio in Montepulciano (1518-1545) by Antonio da Sangallo the Elder, which presents two detached bell-towers in the corners and a lower semi-circular body, the sacristy, joined to the arm where the altar is.

Type B could be a perfect building *in se*, but we detect its presence more as a spatial unit that is part of a more complex composition, such as the plan for an octagonal church presented in MS 2307, fol. 5v (see fig. 1). It is also very similar to the church of San Sepolcro, on which Leonardo and Bramante probably worked, putting forth a proposal for its renovation (fig. 22).

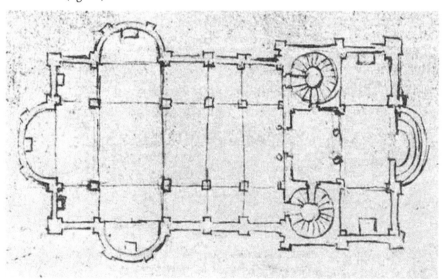

Fig. 22. Leonardo da Vinci, plan of San Sepolcro in Milan. Manuscript B, Institut de France, Paris

With the aid of an interior perspective drawing (see fig. 15), it is possible to make a sure interpretation of the configuration of this space. It is clear that the round dome is now circumscribed by the square defined by the four piers and its articulation is identical to type A. Over the smaller squares the situation is probably repeated, as we can see lanterns that appear on the roof. That is confirmed in the plans and interior perspectives of *Codex Atlanticus* fol. 37v-a (see fig. 16). But it is interesting to note the two solutions advanced by Leonardo to define the limits of these corners: with entablatures (the option for type B), and with arches.

Type B' appears in the same sheet. Here the space corresponding to the squares in the corner is lower and thus the arched vault in between become apparent. Consequently, the articulation of the volumes underlines the subjacent Greek-cross structure of the plan.

Type C and C' both have a semi-regular octagon inscribed in the square. Taking our grid as reference, its larger sides measures two units, the same as the diameter of the semi-circular apses, and the smaller sides √2. Once again, both types could be either independent buildings or spatial units, but we can detect their presence in the Cathedral of Pavia as well as in the Church of San Lorenzo in Milan. The importance of this Early Christian temple is well documented in Leonardo's research on centrally-planned churches. There is even a drawing for a church that is a clear development of San Lorenzo's plan that was executed at the Cathedral of Pavia (fig. 23).

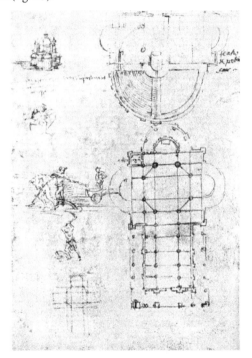

Fig. 23. Leonardo da Vinci, plan for a church based in San Lorenzo in Milan.
Manuscript B, Institut de France, Paris

Type C' with an octagonal dome is the traditional solution, derived from the Gothic period, for covering a prismatic octagonal space, and this is the solution that was employed in Pavia (see fig. 14). Here we can verify that the height of the dome has been reduced in comparison to many of Leonardo's drawings, where the complete veneration of Brunelleschi and the influence of the dome for Santa Maria del Fiore are evident.

Type C, with a hemispherical dome, is the modern response. We know how Bramante solved the transition from the octagon to the circle where the cylindrical dome sits. It is in his project for San Pietro and it is clearly described in a drawing by one of his collaborators (fig. 24). A few years later, Raphael did the same thing on a much smaller scale in the Chigi

Chapel of Santa Maria del Popolo in Rome. We cannot be sure if Leonardo was the first to conceive this solution. His innovative representational system is not sufficient to allow us to be sure, or perhaps the problem is that insufficient information is given in the plan. In this particular case, since we do not have a wooden model, an orthographic section or an interior perspective would be welcome.

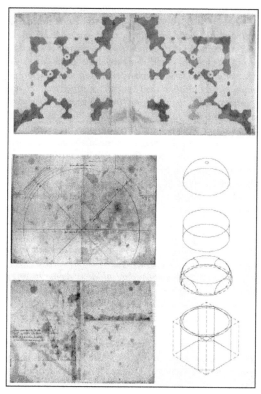

Fig. 24. Design for San Pietro in Rome (1505). Plan by Bramante. Plan and section of the project for the spherical penditives by Antonio di Pellegrino (for Bramante). Axonometry of the system by the author

Peruzzi solved the problem through an original synthesis of both procedures (see fig. 13). But he could not know that his approach was leading him away from the world of perspective and opening the doors for the definitive entrance of axonometry in the field of architectural representation.

Acknowledgment

The author wishes to thank Architect Domingos Tavares, History of Modern Architecture Lecturer and Professor at FAUP, for his advice on this paper, Architect Ana Ferreira, who constructed the virtual 3D models, and Architect Kim Williams, who so carefully revised the text. I am also grateful to Architect Sylvie Duvernoy, guest editor of this issue of the *Nexus Network Journal*.

Notes

1. Perhaps we should add "considerable" above HL presuming an observation from a high point of view.

2. Pietro Accolti, *Lo inganno degli occhi* (1625) relates parallel projection with the shadow projected by the Sun. Directing his attention to practicing painters he states:

   insegnandoci il testimonio del senso visivo (al quale unicamente sottoposta la pittura) manda l'ombre sue, parallele sul piano... con la infinita distanza del luminoso degli opachi... così restiamo capaci potersi all'occhio nostro, in disegnar far rappresentazione di quella precisa veduta di qualsivoglia dato corpo, esposto all'occhio (per così dire) del Sole quale ad esso Sole gli si rappresenta in veduta: onde si come speculando intendiamo il Sole non vedere giammai alcuna ombra degl'opachi, e superficie, ch'egli rimiri e illustri, così tutte quelle, che vengono in sua veduta, intendiamo restare lumeggiate e per contrario tutte le altre a lui ascose restare ombreggiate... Così intendiamo dover essere il suddetto disegno, per rappresentazione di veduta del Sole, terminato con linee, e lati paralleli, non occorrenti a punto alcuno di Prospettiva [Scolari 1984: 46].

3. It wouldn't be fair not to point to Francesco di Giorgio's very intelligent way of drawing. In fact, his perspectives have a great flexibility in order to show at once what has to be shown. Thus he has no problem with intentionally distorting them on behalf of immediate legibility.

4. Among the drawings of *Codice Coner* it is possible to find some pre-cavalier perspectives, especially some architectural details seen from a low point of view (the so-called worm's-eye views, in contrast to the high position of the observer in the bird's-eye view).

5. In the most usual form of cavalier perspective the vanishing angle for the y-axis is 45°; the reduction coefficient along the y-axis is ½.

6. *Non pensi alcuno in queste mie opera vedere mode o regole di prospettiva, l' una per non essere professione di soldato non le saprei fare; l'altra perché li scorci che vi andrebbono, l'huomo leverebbe troppo dalle piante; però in esse piante, e profili consisterà il tutto di queste opera e questa si dirà prospettiva soldatesca* [Maggi and Castriotto 1564], cited in [Scolari 1984: 43].

7. *A groundplan is made by the proper successive use of compasses and rule, through which we get outlines for the plane surfaces of buildings* [Vitruvius 1960: I, ii, 15].

8. Carlo Pedretti speaks about a renewal program of possible Bramantesque inspiration. Cf. [Pedretti 1981: 24].

9. I am referring to perspective where the central vanishing point acts as the unique vanishing point. So, the y-axis is perpendicular to the picture plane.

10. The Latin-cross plan does not raise any new question, since in this situation the nave works as an antechamber to a centrally-planned temple.

References

BENEVOLO, Leonardo. 1993. *Storia dell' Architettura del Rinascimento* (1 ed. 1968). Bari: Editori Laterza.

BOIS, Yves-Alain. 1981. Metamorphosis of Axonometry. *Daidalos* 1: 41-58.

CABEZAS, Lino. 2002. Las máquinas de dibujar. Entre el mito de la visión objetiva y la ciencia de la representación. Pp. 83-348 in *Máquinas y herramientas de Dibujo*. Juan José Gómez Molina, ed. Madrid: Ediciones Cátedra.

COLQUHOUN, Alan. 1992. Assonometria: primitivi e moderni. Pp. 12-23 in, *Alberto Sartoris. Novanta gioielli*. Eds. Abriani, Alberti/Gubler, Jacques. Milan: Mazzotta.

GADOL, Joan. 1969. *Leon Battista Alberti: Universal Man of the Early Renaissance*. Chicago: University of Chicago Press.

LOTZ, Wolfgang. 1997. La rappresentazione degli interni nei disegni architettonici del Rinascimento. In: *L'Architettura del Rinascimento*. Milan: Electra Editrice.

MAGGI, G. and J. CASTRIOTTO. 1564. *Della fortificazione delle città*. Venice.

MILLON, Henry and Vittorio M. LAMPUGNANI. 1994. *Rinascimento da Brunelleschi a Michelangelo. La rappresentazione dell'Architecttura*. Milan: Bompiani.

MURRAY, Peter. 1978. *Renaissance Architecture*. Milan: Electa Editrice.

MURTINHO, Vitor M. Bairrada. 2001. La Piú Grassa Minerva'. A Representação do Lugar. Ph.D. dissertation, University of Coimbra.

MUSEO NAZIONALE DELLA SCIENZA E DELLA TECNOLOGIA "LEONARDO DA VINCI". http://www.museoscienza.it/leonardo/.

PEDRETTI, Carlo. 1981. *Leonardo Architetto*. Milan: Gruppo Editoriale Electa.

RICHTER, Jean Paul. 1970. *The Notebooks of Leonardo da Vinci*. New York: Dover Publications.

SAINZ, Jorge. 1990. *El dibujo de arquitectura. Teoría e historia de un lenguaje gráfico*. Madrid: Editorial Nerea, S.A.

SCOLARI, Massimo. 1984. Elementi per una storia dell' assonometri. *Casabella* 500 (March 1984): 42-48.

VELTMAN, Kim H. 1986. *Studies on Leonardo da Vinci. I – Linear Perspective and the Visual Dimensions of Science and Art*. Munich: Deutschen Kunstverlag.

VITRUVIUS, Marco Pollio. 1960. *The Ten Books on Architecture*. Trans. Morris Hicky Morgan. New York: Dover Publications.

WITTKOWER, Rudolf. 1971. *Architectural Principles of in the Age of Humanism*. New York: W.W. Norton.

XAVIER, João Pedro. 1997. *Perspectiva, perspectiva acelerada e contraperspectiva*. Porto: FAUP Publicações.

About the author

João Pedro Xavier is an architect and geometry teacher. He received his degree in Architecture from the Faculty of Architecture of the University of Porto (FAUP), a Ph.D. in Architecture in 2005, and has been licensed as an architect at the College of Architects in Porto since 1986. He has won academic prizes, including the Prémio Florêncio de Carvalho and Prémio Engº António de Almeida. He worked in Álvaro Siza's office from 1986 to 1999. At the same time, he established his own practice as an architect. He has been teaching geometry since 1985 at the Architecture School of Cooperativa Árvore in Porto, the Fine Arts School of Porto and at FAUP from 1991 to the present. He is the author of *Perspectiva, perspectiva acelerada e contraperspectiva* (FAUP Publicações, 1997) and *Sobre as origens da perspectiva em Portugal* (FAUP Publicações, 2006). Xavier has always been interested in the relationship between architecture and mathematics, especially geometry. He published several works and papers on the subject, presented conferences and lectures and taught courses to high school teachers. He presented "António Rodrigues, a Portuguese architect with a scientific inclination" at the Nexus 2002 conference in Óbidos, Portugal.

Vesna Petresin Robert

4H, Nottingham Street
London W1U 5EQ,
UK
vesna@rubedo.co.uk

Keywords: Leonardo da
Vinci, structural design,
principles of ornamentation,
visual perception, aesthetic
order, ambiguity, optical
illusion, ornament and
structure, pattern, reciprocal
grid, tensegrity, emergence,
Joseph Albers, M.C. Escher,
Cecil Balmond, symmetry

Research

Perception of Order and Ambiguity in Leonardo's Design Concepts

Abstract. Leonardo da Vinci used geometry to give his design concepts both structural and visual balance. The paper examines aesthetic order in Leonardo's structural design, and reflects on his belief in analogy between structure and anatomy.

Leonardo's drawings of grids and roof systems are generated from processes best known from ornamentation and can be developed into spatial structures assembled from loose elements with no need for binding elements. His architectural plans are patterns based on principles of tessellation, tiling and recursion, also characteristic of the reversible, ambiguous structures which led to Leonardo's further inventions in structural and mechanical design as well as dynamic representations of space in his painting.

In recent times, the ambiguous structures in the art of Joseph Albers, the reversible and impossible structures of M. C. Escher, the recurring patterns and spherical geometry of Buckminster Fuller and the reciprocal grids in structural design of Cecil Balmond display a similar interest. Computer models and animations have been used to simulate processes of perceiving and creating ambiguity in structures.

1 Introduction. Leonardo – Architect?

To recognise the extraordinary work of Leonardo da Vinci as architecturally relevant, his design should be evaluated through criteria of aesthetic value, programmatic and functional issues as well as the building performance. This could turn out to be a difficult task: in an absence of any built work, referring to Leonardo as an architect is not self-evident. Furthermore, his notebook *Codex Atlanticus* remained inaccessible for centuries and the influences he may have had on architecture from the Renaissance to present days is difficult to establish. As this fascinating personality also seems to have been disconnected from the majority of his academic contemporaries, his work has often been described as lacking solid argumentation; it is not surprising that besides being praised as a universal genius, the phenomenon 'Leonardo' has also been criticised as a mere scholarly obsession, continuously reinvented as a mythical archetype.

To consider Leonardo as an architect, his work should be studied in a broader context, including his drawings as well as writings, scientific explorations, inventions and paintings. Understanding his design methods and thinking could help reveal the complexity of issues he addressed as well as analogies with contemporary theory and practice.

Leonardo's biography reveals that besides being a brilliant innovator, he had more than a casual interest in architecture and engineering. An apprentice to the Florentine artist Andrea del Verrocchio, he received training in painting and sculpting, and gained excellent technical and mechanical skills. But it wasn't until his emigration to Milan in 1483 that this member of the Florentine painter's guild also began working on architectural and engineering projects and consulting for the Sforza court. Later, the multitude of his talents and interests were given support by the French King Francis I, with whom Leonardo remained appointed as the First painter, architect and mechanic until his death in 1519.

Nexus Network Journal 10 (2008) 101-128   NEXUS NETWORK JOURNAL – VOL. 10, NO. 1, 2008 **101**
1590-5896/08/010101-28 DOI 10.1007/ s00004-007-0058-6
© 2008 Kim Williams Books, Turin

The strength of Leonardo's architecture might lie in innovative conceptual thinking rather than excellence in practice. However, as some of his later studies of plans of cathedrals show, he also studied some of the very practical issues, such as the varying strengths of architectural elements (pillars, beams and arches) and invented a range of building tools. His plans and sketches display a tendency to mathematical perfection of structures in which every component part has an exactly determined position and performance – a concept appearing much later as Kenneth Snelson and Richard Buckminster Fuller's 'Tensegrity'.

Rather than examining the exciting possibility of reproducing Leonardo's architectural sketches as built structures, let us first try to analyse his conceptual drawings of patterns, design objects, architectural and engineering solutions. They offer insight into his understanding of order in both nature and design as dynamic and evolutionary. Analogies with some of the contemporary art and architecture as well as theories in perception and aesthetics can be established.

Leonardo's creative approach to problem-solving was unique in his commitment to test knowledge through experience and to rely on the senses to clarify experience. His willingness to embrace ambiguity, uncertainty and paradox clearly supported his innovative thinking. Leonardo's search for analogies between art and nature, between art and science, and between evidence and imagination allows us to seek analogies when researching the continuity of Leonardo's design concepts and the impact they may have had on modern and contemporary theory and design.

2 Leonardo's views on philosophy and aesthetics

The sense of balance and harmony in aesthetics can be analysed through numeric and geometric order. In the conceptual drawings Leonardo created for architecture and engineering, the use of geometry aims to provide mathematical perfection as well as structural stability. Both of these qualities imply visual stability, inducing a sense of balance and harmony.

But before examining the underlying aesthetic order of Leonardo's design concepts, let us first consider his philosophical background.

As much as Leonardo prided himself on not being a "learned ignorant" but a self-taught man, certain influences on his thinking can nevertheless be established. It seems that it was particularly by reading Cusanus and Alberti that he formed his views on the Universe and its laws, as well as of the correlation between mathematics (geometry in particular) and aesthetics.

As Dr. Tine Germ points out [1999], it was indeed Nicholas of Cusa (Nicolaus Cusanus, 1401-64), German humanist, scientist, philosopher, statesman and cardinal of the Roman Catholic Church who influenced the work of Leonardo. Leonardo seems to have read Cusanus and shared his views on humanism as well as on mathematics.

Mathematics is perceived by both men as the science that allows artists to understand beauty, as Leonardo confirms in the beginning of his *Treatise on painting* [1956]; it identifies harmonic relations and proportions between parts as primary criteria of beauty. In fact, beauty becomes a universal principle characterising nature, art and geometry.

In his work *De Coniecturis II* [Nicholas of Cusa 1972, vol. III], Cusanus presents Man as microcosm inscribed in a circle and a square – an understanding of the human body that goes beyond the medieval theories of Man and the Universe. Similarly, *De docta ignorantia III* [Nicholas of Cusa 1972, vol. I] relates the individuality of every human being as well as his cosmic definition to the concepts of circle and square.

By using the geometrical shapes of the square and the circle, Cusanus illustrates human individuality and uniqueness [Nicholas of Cusa 1972: III, I, 428]. In *De beryllo VI*, man is referred to as *figura mundi* as well as *mundus parvus* [Nicholas of Cusa 1972, vol. XI], a symbolic figure and an anatomical study at the same time, clearly influencing Leonardo's 'Study of human proportions' (the Vitruvian man).

Leonardo's interpretation of Vitruvius's concept of man is both a proportion study and a manifesto on symbolic correlation of man to his body parts, the parts among themselves, as well as man to the universe. *...e l'uomo è modello dello mundo*, he writes [Richter and Richter 1939: II, 242]. The square manifests the divine nature in its limited form – Leonardo explains it as human nature; the circle is the symbol of infinity, which represents absolute truth; to Leonardo, both have the same essence.

Other Renaissance architects also considered the human being to as measure of all things. However, as Wittkower states, this is not an evidence of an anthropocentric understanding of the world [Wittkower 1988; Steadman 1979: 17]. As man was made in the image of God, the proportions exemplified in the human form reflect both the divine and cosmic order. Such a concept of proportions can also be found in drawings by Francesco di Giorgio in his *Trattato di architettura*.

This viewpoint is also reflected in the nine solutions for the tiburio of the Milan cathedral (which was also his most complete project), which Leonardo designed as a "doctor-architect", treating the body of the building and its illness. The notion of building as a body must have been derived from the concept describing the universe as consisting of self-similar units with its characteristics reflected in all its component parts.

Cusanus was the first neo-Platonist to determine philosophically the newly-defined role of an artist as colliding with the notion of a philosopher. His work entitled *De docta ignorantia* is echoed Leonardo's attitude towards the academics: he may have been a great thinker, but Leonardo preferred to mock the learned and rely on his own experience and intuition. In his belief, a true artist was necessarily also a philosopher, because philosophy and arts both deal with nature and its truth. It is through art that the causes of natural phenomena and their laws can be explained and illustrated [Leonardo 1907: 54].

Following his frequent research in dynamics and mechanics of water, many of his drawings of physical phenomena, such as vortexes and their structure, remained present in his attitude towards the aesthetic order in patterns and ornaments. A similar thought appears in the architectural theories of the 1990s by authors such as Greg Lynn [1998], who draw inspiration for architectural form generation from fluid dynamics.

3 Order as a dynamic concept in nature and art

3.1. Order and ambiguity. Leonardo's close observations of natural phenomena led him to an understanding of the dynamics of order and chaos in nature. His artistic and technological conceptualisation reflects the evolutionary principle of cyclical growth and destruction.

Many theories refer to order as a dynamic rather than a static concept, a unity of opposites producing balance. In Cusanus, God – the supreme harmony – is described as *coincidentia oppositorum*. Cusanus also strongly influenced Alberti's concept of beauty based on numeric harmony of proportions of micro- and macrocosm.

Leonardo's attempt to define the inherent order in natural structures in order to understand the macrocosmic level of universal logic is echoed in some of the contemporary theories addressing the evolutionary principle of growth through unifying elements in opposition. The principle of analogies in micro- and macrocosm also appears in contemporary theories of emergence and chaos having their roots in natural sciences.

Order is defined as the state of a structure in a dynamic system with an inherent latent chaos; intrinsic influences can trigger a dynamic imbalance which leads to a chaotic state of the structure. Once such a state has been reached, the structure's tendency is to return to the initial state of order. Therefore, order can be described as a shifting state among chaotic formations.

A concept that is nearly synonymous with order, harmony and proportion of structural modules, is symmetry. It indicates an array of elements in relation to a central axis or a point, or a characteristic of an object split into two identical parts by an axis or a plane. Visually, it signifies balance and harmony as well as matching dimensions of modular elements.

In the introduction to Vitruvius's *De Architectura Libri X*, Daniele Barbaro [1556] explains symmetry as the beauty of order, a harmonious relation of a part to the whole where the matching measures indicate modularity.

However, symmetry can also be defined as a hidden dynamic: the difference between the given structure and one that has undergone a symmetric transformation (for example, mirroring, cyclical rotation, glide plane reflection) cannot be perceived because the structural modules are in balance. The result of such a transformation shows no obvious changes in basic quantities of the structure. Therefore, a structure is symmetrical if it revolves onto itself after a transformation. Natural configurations (for example, crystalline chemical structures) are symmetrical.

Self-similarity is the characteristic of fractal scale: any natural or artificial structure, in which every part is a copy of a part of the structure as well as a copy of the entire structure, is self-similar. Any symmetry that does not change as a result of change in size, is self-similar: smaller parts mimic bigger parts that are similar to the entire structure.

Natural systems behave unpredictably. In fact, nature appears to be a chaotic system where order is present in its smallest components, which grow algorithmically and sum up intrinsic variations, ultimately resulting in a high degree of complexity. Such dynamic, non-linear systems self-organise up to a critical point when an intrinsic event provokes a chain reaction that leads to chaos, that is, where the feedback of those intrinsic events adds up to a drastic change in the system. Complex dynamic systems are said to be composed of interactive elements and are very sensitive to intrinsic factors.

The building blocks of such systems are fractals – geometric shapes whose dimensions have a non-integer value, relating to the degree of folding the particular pattern or shape has. An algorithm based on a geometric motif can evolve into a fractal – a self-similar structure that offers infinite irregularity at any fixed scale. Fractals grow into diverse complex structures following simple iterative equations. Modular systems in architecture have a similar principle: basic modules grow into larger structures by multiplication of modules according to a simple algorithmic rule.

Like modular systems, the Fibonacci series is in fact a scale of self-similar relations in which the values of fractions among the subsequent members of the series approach the golden ratio. The Fibonacci series gives insight into growth by accumulation. Perhaps it is not surprising to find it in both organic nature as phyllotaxis, and in art and architecture as a blueprint for harmony.

3.2. Leonardo: learning from nature. Using his ability to see analogies, Leonardo studied order in nature as a system to be reproduced in his creative work. The use of geometry in his design concepts is aimed at providing mathematical perfection and structural stability, but one could assume that it also provides visual stability and therefore aesthetic harmony, which will be discussed in the following sections. Leonardo wrote a book on the elementary theory of mechanics, published in Milan around 1498 [MS 8937, Biblioteca Nacional, Madrid], in which he warns of the dangers of facing confusion, should one not know of the supreme certitude of the mathematics. His main influences were probably the geometric nature of Cusanus's philosophy, Alberti's *De Re Aedificatoria*, as well as his façade motives and decorative patterns, particularly Santa Maria Novella. Other important influences include Piero della Francesca's *On Perspective in Painting*, Luca Pacioli's *Divina Proportione* (which Leonardo illustrated) and the geometric nature of Islamic patterns and ornamentation, brought to Renaissance Italy through a vivid commercial and cultural exchange.

Cusanus claims that artists mimic nature, but without copying its shapes, forms and visual patterns. It is the spirit that guides and determines the artist. Empirical and sensual experience as well as the underlying order in natural forms is fundamental to the process of perception and cognition: *La natura è piena d'infinite ragioni che non furono mai in isperienza* [Richter and Richter 1939: II, 240] and *Nessuno effetto è in natura senza ragione; intendi la ragione e non ti bisogna esperienzia*" [Richter and Richter 1939: II, 239].

Leonardo similarly described nature as the *maestra dei maestri* [Richter and Richter 1939: I, 372]; he claims a work of art is best when it approaches nature. Studying anatomy was therefore of great importance for his solutions in art and design.

Leonardo worked empirically and studied the impact that nature has on direct sensual experience. Yet in his *Treatise on painting* [1956], he admits that the aim of creation is the artist's vision – a vision that is always already inherent in the Universe. He writes:

> *Sel pittor vol vedere bellezze, che lo innamorino, egli n'è signore di generarle,... et in effetto, ciò, ch'è nel' universo per essential, presentia o immaginatione, esso l'ha prima nella mente, e poi nelle mani.*

> (if a painter wants to see beauty that enchants him, he is the master to generate it, ... and that, which in universe exists as an essence, a presence or

an imagination, first exists in his mind and later in his hands) [Leonardo 1956, fol. 5r, p. 24],

a thought that was repeatedly stated by Cusanus: ... *nam omnes configurations sive in arte statuaria aut pictoria aut fabrili absque mente fieri nequeunt. Sed mens est, quae omnia terminat* (...no artefact in sculpture or painting or craft can be created without the mind. The mind is that, which determinates everything) [Cusanus ---- : VIII, 534].

Leonardo described the universe as a mechanical entity; this is reflected in the way he designs structures and deals with forces, resulting in engineering solutions very unlike the static solutions of his contemporaries. Leonardo's inventions, according to Buckminster Fuller, reflect a universe that is dynamic in every aspect.

In his quest to discover a universal order and the underlying principles of micro- and macrocosmic harmony, the synthesis of anatomy and mechanical engineering is probably Leonardo's most powerful concept.

All his design concepts show an underlying organic order, a synthesis of the natural and the man-made, of structural logic and aesthetic sensibility. The drawings of the manuscript Windsor RL 12608r reveal his simultaneous research in architectural details and anatomic dissecting. In many of the drawings in *Codex Atlanticus*, machines and engineering devices are conceived in imitation of biological organisms and their behaviours. This translation of movements of the organic to the machine world reveals his attempt to rationalise forces in nature through the use of mathematics.

This approach resonates in Buckminster Fuller's interest in biological forms as representations of the "'microcosm in the macrocosm"; it is also reflected in his structural design, such as the faceted shape of the Expo sphere; this structure mimics a marine micro-organism first described in D'Arcy Thompson's *On Growth and Form* [1992] and admired by Umberto Eco because of its ambiguous character [1967: 9].

Engineer Cecil Balmond similarly argues that by looking at Nature, we discover a pattern-maker of infinite skill. The simple and the complex inform each other in a collective exchange where their hierarchies are lost in an ultimate loop. Nature is also information, its algorithms a combination of iteration and a tendency for coherence.

4 Viewing and seeing

4.1. Leonardo's theories of vision. One of Leonardo's principal drivers of creativity was the sharpening and perfecting of his senses to enhance his empirical demonstrations, which explains his obsessive research in anatomy. Following the total solar eclipse on 16 March 1485, Leonardo began not only to study the human eye, but also to design optical instruments, as he was convinced that the human eye errs less than the human spirit. Exploring human vision helped him understand the moving nature of light, and led him to the discovery of its proper speed which he tried to calculate. His treatises on optics focus on the anatomy and functions of the human eye; they can be regarded as attempts to understand the mechanisms of visual perception. As much as they seem remote from today's understanding of vision, they were successfully applied to his experimental and creative work.

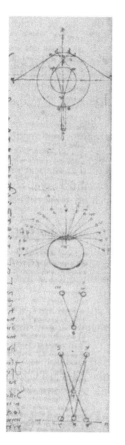 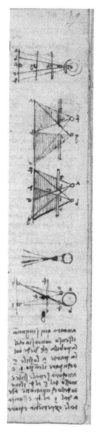  

| a) Studies of the way in which the eye sees objects. Ms. D, fol. 8v | b) Studies of the mechanism of optical illusions. Ms. D, fol. 9r | c) Reception of the image in the eye and schema of the intersection of visual rays. Ms. D, fol. 8r | d) Studies of luminous rays produced not by the object but by the pupil. Ms. D, fol. 9v |

Fig. 1. Leonardo da Vinci's studies on optics

According to David C. Lindberg [1976], the foundation of Leonardo's theory of vision is his theory of radiation and the radiant pyramid. Objects send their images or likenesses (*similitudini*, or species) in all directions into the surrounding transparent medium. The species converge along straight lines and form pyramids that have their base on the object and their apex at every point in the medium. This way, Leonardo believes, each body by itself alone fills the air around it with images; this air receives the species of bodies that inhabit it. It is interesting to note that here again Leonardo states that species are self-similar: all throughout the whole and all in each smallest part; each in all and all in the part.

In Leonardo's view on perspective, all objects transmit their images (species) to the eye by a pyramid of lines which start from the edges of bodies' surface; converging from a distance, they are joined in a single point situated in the eye, which is the universal judge of all objects (fig. 1).

The images spread out in a circular way from an object – as in the case of circles spreading out from the point of impact of a stone thrown into a volume of water.

Leonardo's great contribution to visual theory was his comparison of the eye to a camera obscura, so that the intersections of rays from a visible object must occur within the pupil. This way an inverted image is formed – unless a second intersection appears, caused by reflection or refraction. But the retina is not considered to be a screen, analogous to the back of a photographic camera, onto which images are projected.

Leonardo's critical observation, supported by experimental demonstration, also made him aware of the variable size of the eye's pupil.

Rejecting the perspectivists such as Brunelleschi and Alberti, he argued that the seat of visual power is not situated in a point and is not at the apex of the visual pyramid. He believed the pupil of the eye has a visual virtue in its entirety and in each of its parts and demonstrates that the eye perceives from more than one point.

Later, he claimed that vision occurs only through rays emanating from a point; only radiation falling anywhere within the pupil would be perceived. The judgment of the visual input supposedly resides in the place where all senses meet. Today, this is known as the visual centre in the cortex where cognitive process begins once the visual input has been processed.

Leonardo's studies and experiments in anatomy allowed him to elaborate his own theories of visual perception and representation, and were fundamental to his further observation of order in nature and art.

Today, theory of visual perception is based on the diffuse reflection of light in the process of perceiving images of the environment. Visual perception is the principal active system for retrieving information from the environment and functions as a process of creating images through our sensorial apparatus.

The human retina was once considered to reproduce reality in a way a lens or a mirror would. Today, research shows that complex information perceived via sensorial stimuli become meaningful after they have been processed in the brain, that is, after having been placed in a context of neurological activities.

In psychology, Gestalt theory, formed in the 1920s by Kafka, Koehler and Wertheimer, had a great impact on art and design for the decades to come. It proved that the hierarchy of image and background is the primary principle of organisation of sensorial stimuli. The image is defined as the dominant, coherent element in a visual field, whereas its background (also called the context) is defined as ambiguous, less ordered, secondary and diffuse. It is not surprising that in visual perception as well as aesthetics, theories are built upon similar relations of hierarchy: a whole is more than a mere sum of its component parts, distinguished from the parts by its formal and semantic quality.

According to Gestalt theory and particularly James J. Gibson, the organisation principles of perception reflect the isomorphic configuration of nervous processes (see, for example, [Gibson and Hagen et. al 1992]. Human mechanism of the organisation of perceptive stimuli as well as the constancy of perception is in direct relation to our physiological needs, allowing order in a system to be spotted immediately.

In certain situations however, a clear distinction between the figure and the background is less evident, such as when viewing ambiguous or so-called 'impossible' structures that correspond to the notion of symmetry break in structures.

4.2. Hyperseeing: reading ambiguity. Referring to symmetry and order as 'latent chaos' indicates that there must be a thin line between the two extremes.

Symmetry in emergent theory indicates a lack of interaction with external forces and the context. A structure with such qualities is sensitive; it is inevitably subject to dynamic forces that produce irregularities which sum up to chaotic behaviour. When this occurs, the latent chaos in a symmetric structure becomes apparent; the so-called symmetry break creates dynamics in the structure and becomes the focus of attention. This irregularity can be a broken sequence or continuity, a visual accent, or any other disturbance in the structure. The tendency, however, is for the behaviour to become regular again; therefore a symmetry break is not considered a loss of organisation, but rather a chance for growth and reorganisation within an open, flexible, adaptable, polymorph system. Symbolically, it brings in a dynamic conflict and the notion of opposition and duality.

Ambiguity is an example of such chaotic behaviour. In visual perception, stability and constancy of a perceived set of visual stimuli can be destroyed when misinterpreted.

When we attempt to understand ambiguous structures, a dynamic, relative, shifting experience of illusion occurs as an error in the process of transmission of information along the system. Ambiguous structures have more than a single possible interpretation and always appear as a shifting sequence of interpretations, that is, in an inverse optics. Deleting redundancy from ambiguous elements enables a correct interpretation.

Gestalt theory explains the multistable perception of ambiguity as the dynamic, oscillating perception of ambiguous structures having two equally dominant perceptive organisations that cannot be semantically resolved. In an ambiguous structure, perception shifts between the figure and the background can be created: if the eyes of the observer remain fixed upon an object, the rest becomes background. If the figures are equally strong both structurally and semantically, the observer's attention shifts and so does the interpretation of a figure (an object) against its context. Optical illusions are examples of such ambiguity in visual perception. They occur when an image lacks a hierarchical differentiation of object from background; such a situation does not provide either sufficient information for the brain to decode a single message, or to recognise both interpretations simultaneously, which results in the oscillation between two options, as mentioned above.

The synergetic model of perception describes such dynamic features through pattern recognition: when attention is saturated, only two prototype ambiguous patterns (constants) are said to remain in our memory: our perception resolves this situation by shifting between the two extreme interpretations.

Ambiguity usually indicates weakly organised visual stimuli; it should be recalled that in Gestalt theory, stimuli are said to be organised by sensorial assimilation following the principles of figure versus background as well as group formation (using methods of organisation such as proximity, similarity, good form, symmetry, enclosure and common destination of stimuli). However, certain principles of visual organisation, such as overlapping or diminishing size with distance, which are also commonly used in perspective construction, can have an impact on our perception of depth, distance and space. It is the phenomenon of the constancy of perception that keeps visual organisations stable. The human capacity to recognise colours, sizes, shapes, and light as constants – regardless of minor contextual influences on the perceived image – means that structurally strong organisations can remain stable, whereas weaker structures dissolve and become ambiguous.

One such ambiguous structure is particularly intriguing for artists: the so-called 'impossible' object is a configuration that appears as a result of an intentionally created ambiguity. This phenomenon is at the intersection of disciplines such as art, psychology and mathematics. Multibars, such as the Penrose 'impossible' tribar, are fascinating: here, a contradiction in our interpretation of the perceived object is noticed but is not dismissed as materially impossible. Our brain tries to interpret it as a three-dimensional object in Euclidean space, with straight edges, instead of interpreting it simply as a two-dimensional object drawn on the paper plane. There is no particular element signalling the perception shift or a break that would trigger the "break-spotters" and "continuity" is reported, making the impossible even more obvious. In the interval between fixations, certain characteristics of the object appear as their opposites (for example, convex becomes concave), which we do not normally experience in real objects. This phenomenon has inspired many of Escher's drawings and watercolours.

In mathematics, the four-dimensional space is referred to as hyperspace. According to Nat Friedman [2001], the method of observation in hyperspace is 'hyperseeing'. In hyperspace, one could hypersee a three-dimensional object completely from one viewpoint. Open forms and topological geometric configurations (knots, for example) are particularly interesting for such observations, and viewing ambiguous structures would benefit from this method as well. To entirely see a n-dimensional configuration from a single viewpoint requires stepping into an $n +1$ dimension. Theoretically, this would allow us to see every point on the configuration, as well as every point within it. This approach to viewing has often been used by artists such as Picasso and Bacon, showing multiple views of a configuration in the same representation, and is an essential tool in architectural design and representation, shifting from plan and section to volume and animated views.

4.3. From plan to volume: architectural abstraction, perception and representation of space. Hyperseeing can also be used to describe the ambiguous nature of many of Leonardo's drawings and paintings.

The mechanism of hyperseeing, optical illusions, linear perspective and depth clues can also be used to produce an illusion of space in the representations of a three-dimensional space on a two-dimensional plane. Illusions of depth and movement are used in painting and architectural representation (for example, perspective) to simulate a two-dimensional plane that opens into higher dimensions. In the same way, a two-dimensional pattern containing ambiguity and illusory space can open up into a three-dimensional structure. Leonardo's conceptual drawings of ornaments and architectural structures in *Codex Atlanticus* frequently rely on this mechanism.

In art and architecture, perspective is used as a geometric tool for selecting viewpoints and structuring space in relation to an idealised body. Brunelleschi and Alberti introduced it as a projective tool as well, and as a conceptual scheme supporting theories of visual perception. Perspective drawings are based on the illusion of depth (that is, the illusory three-dimensional space) and overlapping to represent space on a two-dimensional plane; on the other hand, axonometric projections enable an overview of an infinite space. Perspective depends on a single viewpoint, whereas the neutral space of an axonometric projection suggests a continuous space filled with elements in motion. The ambiguity and reversibility of axonometric space allow representation with multiple viewpoints. Unlike the case of perspective viewing, the observer focuses on an object in an indefinite spatial-temporal field.

In Leonardo's representations of architecture, the relation between plan and volume is particularly interesting. His bird's eye perspective drawings of churches indicate his attempt to step away from the limitations of a single viewpoint by using a nearly isometric projection. Being a visionary, he applies spatial parameters that could not be measured or empirically verified with instruments of his time. His landscapes are created using a series of views departing from a single viewpoint.

The landscape in the background of Leonardo's painting of the Mona Lisa, for example, has multiple horizons, with details so accurate that they defy the logic of the linear perspective. The background consists of several plans that were painted sequentially and can also be viewed simultaneously. The entire set appears as if shot with multiple film cameras. Adding to the dynamics and ambiguity of the background, the portrait itself reveals an ambiguity in the shifting facial expressions of the portraitee, achieved with the help of the blurred outlines and colours of the sfumato technique.

Leonardo's use of spatial elements and multiple horizons is also close to the technique used by the German master Albrecht Altdorfer (1480-1535). His multiperspective approach to landscape painting in his canvas *The Battle of Alexander* from 1529 (Alte Pinakothek, Munich) allows viewing of a multitude of details from a single point on the horizon. It gives the viewer the impression of looking at the set from a great distance (as if from high in the air), yet being present at the same time in the mass of bodies, both human and animal, at all the key viewpoints. The painting can be interpreted as having an omnipresent observer situated both in the sky and on the ground.

Conventional architectural representation tends to frame a moment in time but cannot represent the forces shaping it and allowing all those subtle shifts between Euclidean dimensions that take place in a dynamic system. Computer animation integrates the dimension of time into architectural representation; it shows an evolutionary environment in which processes and their forces shape the built structure.

5 Dynamic principles in Leonardo's design

The timelessness of Leonardo's inventions is reflected in the way he adopted universal principles of mechanics and anatomy to create dynamic structures. The design methods used in his sketches and in his graphic, ornamental, structural and mechanical designs were inspired by his investigations in morphology.

5.1. Dynamic order: rhythm, sequence, pattern. The discourse of architectural representation and composition has benefited from the continuous research of mechanisms of human perception. Design concepts can be developed based on rules of visual perception; vice versa, certain principles of visual organisation (for example, overlapping and diminution of size with distance) have been demonstrated to influence our three-dimensional perception strongly.

Visual perception in art and architecture is considered a variable, but geometry on the other hand provides empirical foundations: it defines relationships among elements of a composition.

In the design vocabulary, symmetric transformations – rhythm, variation, number, measure, proportion, periodicity and modularity, repetition, as well as translation, rotation, and reflection – are used to generate patterns or spatial concepts.

Human creativity relies in large part on our ability to recognize patterns, which can be compared and matched with ones previously memorised. These patterns can manifest as different types of sensory output, making analogies among disciplines possible.

The elementary means of creating patterns and a sense of order in a structure is the process of repetition. According to Matila Ghyka [1971], rhythm can be described as observed and recorded periodicity. Periodical repetition of a motif or a pattern introduces a sense of movement and dynamics (that is, the dimension of time) into structure. Rhythm is one of the basic aspects of energy in time, creating patterns and sequences of form with specific frequency. It emerges from iteration and variation, as well as contrast, such as figure versus background.

In any rhythmical sequence, a pattern is produced when a motif appears. A very regular pattern can be quickly perceived because of the way visual organisation functions: irregularities such as ambiguity are easily spotted since, from an evolutionary viewpoint, they indicate change and possible danger.

5.2. The dynamic order of Leonardo's lattice grid. Examining the aesthetic order in Leonardo's patterns, ornaments and design concepts, particularly the plan of a roof system as appearing in *Codex Atlanticus* gives insight to the nature of perception of order and ambiguity as Leonardo saw it (fig. 2).

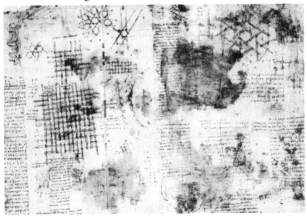

Fig. 2. Leonardo da Vinci, study of a wooden roofing made of parts that fit together. *Codex Atlanticus* fol. 899 v

The rectangular lattice structure appears in *Codex Atlanticus* fol 899 v as a study of wooden roofing made of parts that fit together. The lattices are tiled using a monotonous repetition with overlapping endpoints rotated at 90°. Overlapping (transposition) is used as a method of transformation that develops periodicity, regularity and rhythm in the pattern. By imitating plaiting and weaving, the pattern of the structure displays modularity and self-similar growth similar to the iterative order found in the generation of emergent structures. In Leonardo's drawings, the iteration of a fixed set of elements (beams) with a recombination (rotation and overlapping) creates a diverse and variable, yet modular structure.

Leonardo uses overlapping to develop a possible dome structure out of a grid. Here, the principle of overlapping has another interesting effect: as later shown by Gestalt theory, certain principles of organisation of visual stimuli in any two-dimensional representation can be used to create the illusion of depth, distance and space. Overlapping is a strong visual clue: it indicates relations among elements that are distributed hierarchically according to depth, that is, their location in an illusionary three-dimensional space. This technique has been extensively used in painting.

Leonardo's sketches demonstrate an aesthetic order derived from compositional processes in ornamentation, plaiting and weaving, featuring compositional methods that involve symmetry. Achieved by applying one or many of the three crystallographic principles, they involve shifting – through translation, rotation, reflection, for example – as well as tiling. The symmetry break, characteristic of ambiguous structures, is frequently used by Leonardo in his multi-perspective approach to landscape and portrait painting mentioned earlier. In Ernst Gombrich's terminology, such dynamics can be described as an exchange of restlessness and repose [1979: 120-126].

Although appearing symmetrical, the lattice structure drawn by Leonardo contains a symmetry break that introduces a dimensional transition from a two-dimensional planar image to a three-dimensional space. A perceptive shift takes place among the recursive elements that make up the structure. An illusion of stepping into the third dimension from a two-dimensional plane appears due to the ambiguity produced by overlapping as well as shifting of the depth clues. This ambiguity is perceived as oscillation between the convex and the concave, with shifting dimensions due to the confusion that the viewer of this drawing senses between the back and the front side of bars. Today, we refer to this phenomenon when describing the two-dimensional projection of a wireframe cube or the unstable perception of a hypercube. The illusion of continuity in which our perception follows a loop without noticing the twists (the symmetry breaks) gives this iterative structure its ambiguous dynamics (fig. 3).

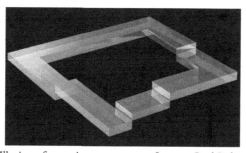

Fig. 3. Illusion of a continuous structure. Laurent-Paul Robert, 2003

In *Codex Atlanticus*, a number of drawings can be found that display designs imitating patterns, knots and plaits, details in wooden structures, the puzzle-like connection details of wood, chain members, intarsia, bridges (such as the wooden bridge on display in Vinci, barriers and birdcages. The design of the growing grid of the roof system may have been derived from Leonardo's interest in problems of continuous interlacing. This research has initiated the publication of a series of knots called 'Academia di Leonardo da Vinci' as a response to complex interlacing patterns adapted by Italian craftsmen from Islamic examples. Similar to textile design techniques, the aesthetic order of the structure is derived from processes used in ornamentation, plaiting and weaving (for example, linking, vaulting, overlapping). The construction of this roof system does not rely on additional binding elements; it is conceived as a self-revolving, self-supporting and continuously evolving structure.

The pattern of the roof system that Leonardo proposed was most probably inspired by studies of knots and puzzle-like structures and textures. Its approach is similar to his many mechanical engineering and defensive devices. Leonardo offers a dynamic solution to a static problem: forces within the structural beams of the dome are distributed using overlapping and linear transformations. He frequently used the principle of grasping physical forces through the rhythm and geometry of structure. His research in mechanical structures and devices, as well as water, revolved around pattern analysis that influenced his design concepts.

From an architectural perspective, the relation between plan and volume in Leonardo's design is particularly interesting. The roof plan is a two-dimensional abstraction of a three-dimensional structure, yet it functions as an ornamental pattern: its ambiguous character creates a visual shift, a seamless transition from plane to space.

Due to the temporary inaccessibility of Leonardo's drawings and an absence of built work, his influence on architecture is considered minimal, yet some of twentieth-century art and architecture displays similar interests.

The principles of growth and expansion from the second to the third dimension using overlapping and self-similarity can be found in many of M. C. Escher's graphic works. Fascinated by the concept of bounded infinity, Escher tried to reproduce growth through iteration.

The triangular variation of Leonardo's periodical pattern of the roof system (as in fig. 2) is based on superimposed overlapping beams that create an ambiguous pattern with two possibilities of interpretation. The pattern produces an illusion known from the previously described perceptive loop of the 'impossible' multibars, particularly the Penrose tribar. Each new situation that the eye encounters when following the lines of the pattern causes a re-adaptation of the interpretation, while the twists are not at all perceived.

Escher achieves the same optical illusion of a continuous structure that follows a perceptive loop (as in fig. 3) in his lithograph *Waterfall* (1961) inspired by the Penrose tribar, in *Ascending and Descending*, and in *Belvedere*. When observing the structure and gliding along the sides of the object (or the falling water), its impossibility, or the mistake, cannot be perceived. But after having completed the cycle of viewing, the shift takes place in the interpretation of distance between the eye and the observed object. Escher explained the impossible triangle as being fitted three times over into the picture. The Penrose tribar illusion contains ambiguity in its triple warp of the structure between convex and concave.

Whether the creation of perceptive shifts in Leonardo's lattice pattern was intentional or not, it is important to note that the subject of ambiguity is addressed too frequently in his work to be random. He studied it systematically, as is evident in the right bottom image on the MS H fols 32r – 33v and fol 35r (fig. 4). Here, a lack of a distinction between figure and background (that is, the front and the back edge and planes of the cubes in the arrangement) creates an inversion in the orientation of the cubes that takes place after a few seconds of focused viewing. An ambiguity of convex/concave in this repetitive structure is created.

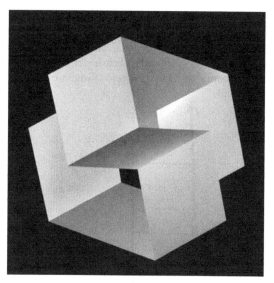

Fig. 4. Leonardo da Vinci, studies of convex-concave ambiguity in an ornamental structure. MS H ff 32r-33v, fol. 53r (detail)

Fig. 5. Structure with a convex-concave ambiguity (an interpretation of a detail from Josef Alberts, *Trotz der Gerade*, 1961). Laurent-Paul Robert, 2003

Due to this uncertainty, the two-dimensional structure depicted introduces an illusionary space drawn in an axonometric projection. The case is similar to the Schroeder illusion with the inversion of a staircase. Even though the visual focus remains, the interpretation shifts from frontal (standing) to backwards (hanging) in a continuous loop.

An analogy can be made with the impossible or irrational cube in the Necker illusion. This impossible object is a two-dimensional drawing of a cube, seen in an axonometric projection with its parallel edges appearing as parallel lines in the drawing. The seemingly consistent figure of a wireframe cube contains ambiguity: its overlapping opposite corners are joined together, resulting in a confusion between the back and the front of the structure. Its top and bottom parts, when viewed separately, appear to be a regular cube. If viewed simultaneously, the image seems to twist because the information is incomplete and lacks the details necessary for a single interpretation.

The space within the image of the cube appears to be warped. This warp is created by half of the cube being rotated 180° and mirrored. The interpretation of the cube oscillates between pointing forwards and backwards.

The same ambiguity is used in *Belvedere* by M. C. Escher in the impossible structure of the tower's upper floor (a character contemplating a Necker cube is also featured). Escher's encounter with the writings of Gestalt psychologists introduced him to the ambiguities inherent in figure-background relationships, the two-to-three dimensional ambiguity on a flat surface, as well as ambiguity of the reversible cube and of multiple perspectives.

Introducing movement into two-dimensional representation of a three-dimensional space through hyperseeing also characterises the work of Josef Albers. In his series *Trotz der Gerade* (Despite the straight line), influenced by Gestalt theory, a perceptive shift between the convex and concave interpretation is produced in the representation of the cubic structures. Albers demonstrates that inversion and permutations that are rich in ambiguities can be achieved with even the simplest elements (fig. 5). The subject of optical illusions, particularly the Necker and Schroeder ambiguities, has also been explored in the *Graphic Condensation* series by Francesco Grignani.

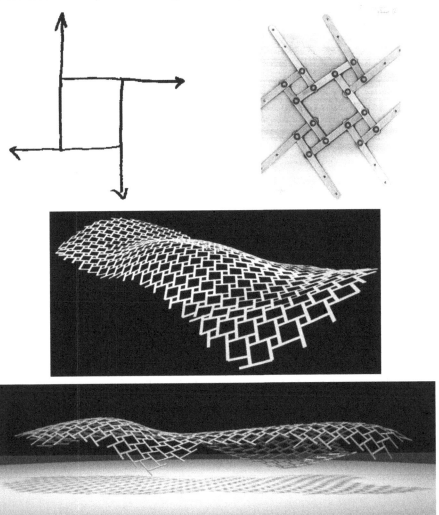

Fig. 6. Structural system: Reciprocal grid shell structural system for Forest Park Pavilion, St Louis, Missouri, USA. Architectural design by Shigeru Ban for Forest Park Forever. Structural concept and integrated design by AGU – Cecil Balmond, Charles Walker, Martin Self, and Benedikt Schleicher First concept phase 2004; scheduled completion 2007. © Sketches and photographs: Cecil Balmond; © Computer renderings: Advanced Geometry Unit – Arup

5.3. Reciprocal grids: a dynamic equilibrium. Leonardo's drawing of the overlapping grid can be translated into a three-dimensional structure assembled from loose, straight elements, when the middle point of one element is connected to the endpoint of another one. This pattern is extended using a simple iterative method, linking and interlacing. The beams represent the linking lines that cross above or below each other and thus produce a three-dimensional arrangement.

The overlapping grid has historical precedents in the spiralling poles of the native North American tepee and the "Rainbow" bridges of China's Song dynasty. Studies by the medieval master builder Villard de Honnecourt also described reciprocal structures that would allow floors and ceilings to be built from short beams.

Leonardo's lattice and triangular grids are in principle reciprocal grids. Elements arranged in a mutually supporting pattern create reciprocal structures. Normally an area is spanned using a grillage or by a hierarchical system of primary and secondary beams. But in the reciprocal grid, a network of individual elements shapes the load-paths in nested loops. The reciprocal structure is less efficient than an equivalent continuous grillage but because no bending loads are transferred between elements, their connections can be very simple.

Balmond argues that, conceptually, using elements that are too short for a continuous transmission of load in a straight line is nevertheless interesting: an interruption, a *staccato* appears in the assembly of the grid. Using a reciprocal arrangement of one member being fed into the other, the grid moves so that a cascade effect is developed.

Working with architect Shigeru Ban, Balmond developed the structural concept and integrated design for the Forest Park Pavilion in St Louis using the overlapping grid principle. Instead of the regular repeating grillages, this reciprocal grid shell has an alternating pattern of large and small squares. Balmond describes it as having a shifting pulse, or as a grid that has two rhythms, where the eye jumps over alternating scales as the passage of axial load is interrupted.

If one edge of the square of the structural diagram is extended, and other sides follow in rotation, then a pinwheel configuration is produced. Materialised as pieces of wood that construct the pattern, each strip is placed on top of the previous one. This way, a weave with different convex curvatures emerges.

If the strip of wood is three units long and the small square has a side of one unit, the unit square will be one-fourth the area of the big square. The area relationship of small to big, and the thickness of each piece of wood as it engages the weave, create different curvatures.

If the system of assembly is flipped over (that is, from an arrangement in which each piece is on top of one another to one in which each one is placed below the other), then the curvature reverses to concave.

In emergent design, a grid may turn into patterns of rhythm, framing multiples of squares, as reciprocal bearing structures arrange themselves through space in alternating plan forms. This way the grid becomes not just Cartesian but Emergent; examples can be found in the concepts Balmond developed for the Taichung Opera House (architect: Toyo Ito), the Fractal Wall of the Grand Egyptian Museum (architect: Heneghan Peng) or the Forest Park Pavilion in St Louis (architect: Shigeru Ban) (fig. 6).

Leonardo's reciprocal grid of the roof system in *Codex Atlanticus* can be used as plan for construction of a dome, a sphere, a cylinder, a column or other structures from loose beams. In his drawings, Leonardo was also testing roof systems as domes based on different geometric shapes, such as the triangle. Buckminster Fuller similarly discovered that if a spherical dome structure was created from triangles, it would have unparalleled strength. Triangulation (fig. 7) allows a minimum number of points on a surface to define a shape that can be stable in space and can describe any complex shape or body. This principle directed Buckminster Fuller's studies toward creating a new structural design, the Geodesic dome, based also upon his idea of "doing more with less"; in this way, the largest volume of interior space can be enclosed with the least amount of surface area, thus saving on materials and cost.

Fig. 7. Structure based on a double Moebius strip with surface triangulation that creates an illusion of depth. Laurent-Paul Robert, Vesna Petresin Robert: Double Moebius Strip Studies, 2002

Buckminster Fuller's geodesic domes, first presented at the 1954 Milan Triennale (entitled "Life Between Artifact and Nature: Design and the Environmental Challenge"), are based on an iterative triangular growing pattern, close to Leonardo's pattern of the roof system. Recurring patterns and spherical geometry allowed Buckminster Fuller to create light, flexible structures such as the 'Tensegrity sphere'.

Tensegrity or "tensional integrity", introduced by Kenneth Snelson, provides the ability of structure to yield increasingly without ultimately breaking. This method of construction is highly efficient and reflects natural generation of structures at cellular levels.

Similar to Leonardo's experiments in frequency and geometry of the roof structure for a dome, Buckminster Fuller's geodesic domes were conceived as fractional (triangular) parts of complete geodesic spheres. A high-frequency dome has more triangular components and is more smoothly curved and sphere-like; but geodesic domes may also be based on other polyhedra, such as the octahedron and tetrahedron. Each of the triangles of the dome subdivision is curved because it is subdivided into smaller triangles, the corners of which are all pushed out to a constant distance from the sphere's centre. This division follows a pattern similar to fractal logic in the growth of space-filling curves or structures that exist between two and three dimensions.

The principle of economy in nature id echoed in modularity. Buckminster Fuller expressed them through his principle of synergetics in a series of prototypes and experimental structures, maps and diagrams.

5.4. Ornament and structure. A sequence of patterns produces ornament that is functionally referred to as decoration. Ornaments can be considered as geometric mandalas and may have structural potential. Leonardo shows that geometric patterns can become blueprints for architectural layouts.

Non-structure-related ornament started disappearing from architecture after Adolf Loos published his 1908 manifesto [1998]. His position was shared by Le Corbusier, who thought decoration was of a sensorial and elementary order, as colour; in his opinion, ornament was suited to simple races, peasants and savages who love to decorate their walls.

But before dismissing ornament as architecturally irrelevant, it should be recalled that Leonardo's drawings reflect the possibility of using its geometric potential to address structural performance in three dimensions. The search for connections at all scales, from microcosmic to macrocosmic, is echoed in his design research, marked by universal principles of geometric, structural and aesthetic equilibrium.

The drawings in *Codex Atlanticus* reveal Leonardo's research in analogies between geometric patterns, their ornamental application and structural potential. His interest in architecture initiated various experiments in engineering and materials. A series of bridge designs were created. Aimed at using minimal profiles, lightweight structure, transportable materials with small profiles and solid, stress-resistant binding elements, these structures were easy to build and, again, followed the interlacing and overlapping principles.

Leonardo uses geometry to achieve stability, order and beauty as he perceived them in the dynamic systems of nature. The dynamics of human perception is reproduced in his structural design. Symmetrical patterns with inherent ambiguity that allow multiple interpretations help Leonardo to adopt universal principles of mechanics, anatomy and bionics.

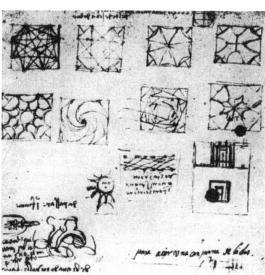

Fig. 8. Leonardo da Vinci, Diagrams of cupolas based on a square plan. MS B fol. 10 v

The evolution of patterns in ornamental design through symmetric transformation (rotation, translation and reflection) is similar to basic operations in architectural morphogenesis. Leonardo applies it to his series of diagrams of cupolas treated as ornaments (fig. 8): he achieves symmetry using the crystallographic principles of translation, rotation and reflection. Translation of an interval produces a rhythmical sequence. In a way, geometric patterns mimic crystalline molecular structure; at a microscopic level, structurally stable shapes also coincide with the elementary geometric shapes indicating visual harmony.

A similar mathematical and crystallographic influence can also be found in the fascinating graphic inventions of Escher, such as in the *Symmetry* watercolour series. Their original inspiration, however, came from his familiarity with psychology and experiments in visual perception. His fascination with order and symmetry seems to have been a later development.

Like Alberti and Leonardo, Escher's inspiration came from the Islamic arts of ceramics, ornamentation, tiling patterns and calligraphy that he discovered at the Alhambra, the abstract geometry and figures merging with each other or the background. He reproduced these principles for a tiling of a surface, using regular plane division as well to create ambiguous structures with perception shifts between figure and background.

Pattern stands for regularity in a particular dimension, that is, repetitive units (motives) ordered by linear (translational) or rotational symmetry. When organised in time, patterns produce rhythm. Repetitive, iterative units indicate a re-use of existing information, producing self-similar structures.

Any organised behaviour can be identified as a pattern within complex natural systems (for example, in physics, biology and chemistry). In art, conceptualisation and construction of patterns in various materials and technologies involves weaving, knitting, printing, matting (as in textiles) rather than basic repetition of motive. These methods can often be found in ornamentation, in graphic as well as architectural design.

Ornaments derived from methods of textile design provided a basis for Leonardo's research of their structural potential: here, again, art and geometry come together. The intriguing research of space-filling patterns to produce the construction grid of an ornament is particularly interesting in Leonardo's drawings of sword heads and ornamentation (fig. 9).

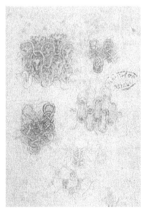

Fig. 9. a, above: Leonardo da Vinci, chain ornament. Codex Atlanticus fol. 681v; b, top right: study of decorative knots and plaiting, Ms. H, H1, c. 33r: c, bottom right: Dismountable chain. Madrid Ms. I fol. 10r (detail)

Periodicity of these compositions is achieved by transformation processes such as translation, rotation and glide reflection, giving a sense of directional movement of the repetitive patterns. Iterative, growing curvilinear patterns mimicking methods in textile design such as interlacing, interweaving and overlapping produce ornament that is analogous to space-filling curves. The drawing technique suggests further ways of development of these ornaments into three-dimensional structures through possible intersections in space.

The iterative growing patterns of 'Academia di Leonardo da Vinci' and *Codex Atlanticus* can be compared to the so-called space-filling curves and monster curves in mathematics. These are generated by iteration, producing a growing configuration that is neither a two-dimensional shape nor a three-dimensional solid. In topological geometry, formations such as a Hilbert curve, knots and the Sierpinski gasket are known to create the ambiguity of evolving from one-manifold (a two-dimensional curve) into two-manifold (a surface) by self-revolving or space-filling pattern growth.

Leonardo's self-revolving, self-similar interlacing ornaments (fig. 10) are close to the notion of fractal dimension (that is, dimension between dimensions) at a very small scale. Emergent aesthetics also offers an analogy to Leonardo's ornamentation in addressing the relations of the whole to its parts: although the whole is seemingly chaotic, an individual part is an ordered, active element generating processes of modifications in time.

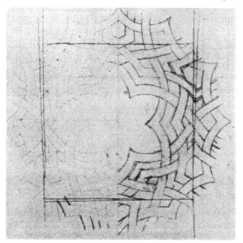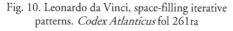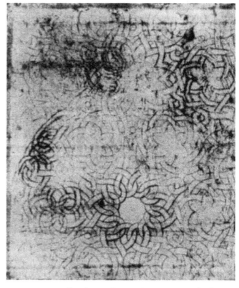

Fig. 10. Leonardo da Vinci, space-filling iterative patterns. *Codex Atlanticus* fol 261ra

Inspired by research in mathematics and physics, especially the spherical concept of the universe, Buckminster Fuller used self-stabilising patterns that are similar to self-similar patterns through which latent order in emergent systems manifests. Form can be retained due to a recurring pattern. Buckminster Fuller therefore defines structure as "locally regenerative pattern integrity" [1975: §606.01]. Such logic of stabilisation and dynamic growth is known in self-similar generation of two-dimensional patterns or three-dimensional structures in a complex system.

Leonardo's attention to details, patterns, geometric ornaments, calligraphy and astronomy is also echoed in the recent research by Cecil Balmond. He explains patterns as building elements of structural networks and as mediators between metaphor and actuality.

The idea that an electronic ornamentation can be created based on computer-generated pattern using the principles of translation, notation and writing, has been developed by dECOi. The lines derived from the rotation of animated forms generate a spatial pattern, as in the electroglyph *Hystera Protera* [1998; Zellner 1999: 69; Goulthorpe 2007: ch. 3]. The computer-generated scriptural sequences are derived from mappings and are referred to as 'glyphics' (to distinguish them from the determinate character of graphics). This fluid, serial, dynamically evolved generative configuration also works as a detail, a structure or a decoration. Electroglyphs are more than a mere multiplication of a series of decorative elements: they can become activated patterns on a responsive surface. In the *Hystera Protera* project, the patterns produce electronic ornamentation that can be generated alternatively by electronic-sensory input or environmental stimuli.

5.5. Ornament as structure and as symbol. The potential of ornament as a structural blueprint and as a semantic entity having symbolic content is fascinating.

An early mythological representation of dynamics found in representations of infinity and eternal revolution is the serpent biting its own tail, such as, for example, the alchemistic ouroboros. Self-revolving patterns and structures were popular, demonstrating an interest in continuous renewal of energy in all manifestations.

Perhaps it is not surprising to find these characteristics in the intertwining spirals of the double helix. The evolving, twisting, dynamic structure of DNA is one of the most stable configurations, based on a generation of an iterative pattern in three-dimensional space. The symbolism of this intertwining, continuously growing structure has also been used as a motive in decorative arts since early times.

In his research in continuous configurations, Leonardo addresses the evolutionary nature of the dual relation between man and universe, between architecture and environment, between outside and inside.

In Pictish and Celtic ornamental art, known to abound in such metaphors, a continuous line was produced using numbers with no common factors with half-sizes at the four corners. Based on methods of textile design such as plaiting, dynamic evolutionary structures, such as knots and spirals, were introduced.

A knot is defined as a curve in space with the ends joined, its two-dimensional diagram showing its crossing points. It is the embedding of one closed curve in three-dimensional space. In mathematics, knots are the subject of investigations of topological geometry along with such self-revolving, continuous phenomena in four-dimensional space as the Klein bottle, Borromean rings and the Moebius strip.

In some of Escher's watercolours, the optical illusion is achieved when the perception of a structure follows a loop. He created a woodcut printed from three blocks in 1963 entitled *Moebius band* that also represents this single infinite surface, always ending up at the starting point, having no beginning and no end.

> An endless ring-shaped band usually has two distinct surfaces, one inside and one outside. Yet on this strip nine red ants crawl after each other and travel the front side as well as the reverse side. Therefore the strip has only one surface [Escher 1992: 12].

In hyperseeing, knots appear to be ideal objects for such a method of viewing, being open rather than solid, with no clear orientation and directional distinction, and appearing different when viewed from various viewpoints.

The structural principles found in knots, plaits and puzzles led Leonardo to design bridges, patterns in wheels, details of wooden structure, chain members, barriers and other devices. He studied the self-revolving as well as growing spirals appearing in vortexes and helical flow diagrams in order to understand principles of rotation, evolution and motion. The drawings in *Codex Atlanticus* (fig. 11) are based on principles of knots – weaving, plaiting, overlapping and twisting – that have been used to create patterns and ornaments. Using transformation processes such as rotation, reflection and translation of a repetitive pattern gives a sense of a directional movement that develops along a curvilinear path. In Leonardo's drawings, such investigations often indicate a possibility of development into structures. Analogies to the organic world are numerous: Leonardo illustrated the principle of ascending spiralling vector that appears in natural phenomena such as vortexes and helical flow in fluids in a series of analytical drawings (fig. 11, above).

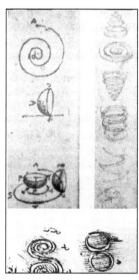
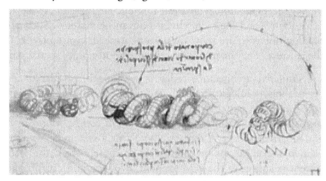

Fig. 11. Leonardo da Vinci. a, left) Studies of knots and spirals. *Codex Atlanticus* fol 292v-a; b, above) Spiral-form mazzocchio formed of two intertwined spiral rings. *Codex Atlanticus* fol. 520r

Similar to Leonardo's drawing of the weaving, twisting ribbons, Escher's woodcut printed from two blocks, entitled *Spirals* (1953), is also an ambiguous self-revolving structure; it has been created using a looping, spiralling structure. Four spiral-shaped bands enclose a twisted tube that revolves to its starting point in a thinning torus-shape; after having completed a tour, the structure makes another tour in a self-revolving way.

In structural design, methods of textile design are used to achieve maximum structural performance of architectural components. Such configurations can adapt to dynamic loads along the three axes of the coordinate system by remaining flexible and elastic. Today, the process of three-dimensional weaving in textile industry is used to create various densities of structural fabric and has been successfully applied to produce resistant structural membranes of great tensile resistance

6 Learning from Leonardo

Leonardo's attempts to conceive a mathematically perfect structure can be compared to the contemporary idea of structural coherence as a way to achieve a dynamic balance. His design is mainly derived from observation of natural behaviours and defies the principles of traditional engineering.

Emergent design similarly relies on the principles of structural growth, efficiency and regeneration found in biological and physical processes. The organic structural principle requires more rather than thicker elements to achieve structural performance. These can be subdivided according to the principle of fractal branching and not in terms of the traditional hierarchy of load-bearing elements. The same thought appears Leonardo's fractal-like subdivision of building components.

Emergent theory has become increasingly important for structural design: emergence is defined as that which is produced by multiple causes but is not directly the sum of its individual effects. Borrowed from biology, mathematics, information theory, artificial intelligence, climatology and bio-mimetic engineering, its key concepts are self-organisation of structures and the hierarchy of bottom-up systems. Such open systems, known from the organic universe, are dynamically imbalanced, constantly in the process of exchange with the context. They adapt by keeping their autonomy, spatial articulation and hierarchies.

Similarly, architecture today is no longer considered an end-product but a process resembling a biological model, constantly optimising according to its environment. Forms and generative patterns are shaped by processes that generate forces of organisation. The principle of adaptation to the environment can be found in Artificial Intelligence systems and smart materials. A dynamic whole or a system of structural patterns is constituted of elements that cannot be interpreted separately from the system. An open, process-based system adapting to the environment, evolving and creating new structures, is the concept common to both the organic world and architecture.

Appreciating Leonardo's notion of the artist who learns from nature, contemporary architecture also investigates geometry of organic shapes to create patterns, ornaments or structures using morphogenetic processes found in genetic algorithms, bio-mimesis and particle physics. The underlying structure of nature is self-similar.

Leonardo's use of geometry in his design concepts is aimed at providing mathematical perfection and structural stability, as well as visual harmony. His architectural concepts reflect an understanding of physical laws, characteristics of materials and self-similarity in organic and built structures, along with a belief in universal principles of mechanics.

The underlying aesthetic order can also double with structural balance. Leonardo must have understood these principles by observation of structures and behaviours in nature. But his interest in concepts rather than construction resulted in a tendency to balance materials and forces in his projects so carefully that any additions to the structure would be unstable; the structure is self-sufficient, with all its elements in precisely defined locations. Such an approach is not far from the tensegrity of Kenneth Snelson and Buckminster Fuller, where every structural member is uniquely located to achieve an integrated optimal performance.

The morphological dynamics of Buckminster Fuller's geodesic domes reveal an ambition similar to Leonardo's: combining structural stability and mathematical perfection with visual harmony. Structure and geometry are interrelated not only in performance, but also in aesthetic requirements.

Evolutionary techniques in architecture involve design principles such as synergetics, tensegrity, organic logic and metaphors in structural design, as well as emergent technologies, to achieve a better structural performance and aesthetic balance (fig. 12). Synergy in design indicates that design concepts involve structural stability as well as flexibility.

Fig. 12. The basic floor plan shape is derived from the logarithmic spiral. Each iterative step rotates a floor plane and translates it along the z-axis. Laurent-Paul Robert and Vesna Petresin Robert: Ophiomorphos, 2002

Synergetics is the principle prevailing in design in imitation of biomorphic structures: tensegrity components require perfect positioning of every single element. The same can be said for both Buckminster Fuller's geodesic domes and Leonardo's carefully calculated structure for the dome. Synergetics is the geometrical coordinate system discovered by Fuller, using the tensegrity principle of interactive relation of parts with the whole. He describes synergy as "the behaviour of whole systems not predictable from the behaviour of separate parts" and the "behaviour of integral, aggregate, whole systems unpredicted by behaviours of any of their components or subassemblies of their components taken separately from the whole" [1979: 101.01-102.00].

Whereas classical and even Modernist architecture polarised ornament and structure, contemporary techniques allow for integration and blurring of inside and outside. A new awareness of surface gives it a role similar to that in biology. In architectural terms, this enables a freedom of expression and form, allowing a diversity of the design addressing patterning and texture, as well as the sensuous and the ornamental.

Leonardo's most important concept in design was the search for analogies between natural and man-made, between aesthetics and functionality. By trying to understand the micro- and macrocosmic harmony, his design successfully resolves not only the issues of ergonomics, but also archetypal human needs – both physiological and psychological – in the built environment.

References

BARBARO, Daniele. 1556. *I Dieci Libri dell'Architettura di Vitruvio Tradutti et Commentati da Mons. Barbaro*. Venice: Francesco Marcolini.

NICHOLAS OF CUSA. 1972. *Nicolai de Cusa Opera Omnia iussu et auctoritate Academiae litterarum heidelbergensis ad codicum fidem edita*. Eds. Josef Koch, Karl Bormann, Hans G. Senger. Hamburg: Felix Meiner Verlag.

DECOI. 1998. *Hystera Protera* Graphic Design / Art Work, Public Art Commissions Agency UK.

ECO, Umberto. 1967. A theory of expositions. *Dot Zero* 4 (Summer 1967).

ESCHER, M. C. 1992. *The Graphic Work*. Köln, Benedikt-Taschen Verlag GmbH.

FRIEDMAN, Nat. 2001. Hyperseeing, Hypersculptures and Space Curves. *VisMath* **3**, 1. http://www.mi.sanu.ac.yu/vismath/friedman/index.

FULLER, R. Buckminster. 1975. *Synergetics. Explorations in the Geometry of Thinking*. 2 vols. New York: Macmillan Publishing Co. Inc.

GERM, Tine. 1999. *Nikolaj Kuzanski in renesancna umetnost: ikonoloske studije*. Ljubljana: SKAM.

GHYKA, Matila. 1971. *Philosophie et mystique du nombre*. Paris : Editions Payot.

GIBSON, James J., HAGEN, M A, et al. 1992. Sensory processes and perception. Pp, 224-281 in *A century of psychology as a science*. Eds. S. Koch and D. Leary. Washington DC: American Psychological Association,

GOMBRICH, Ernst. 1979. *Sense of Order. A Study in the Psychology of Decorative Art*. London: Phaidon Press Ltd.

GOULTHORPE, Mark. 2007. *The Possibility of (an) Architecture. Collected Essays by Mark Goulthorpe, dECOi Architects*. London: Routledge / Taylor & Francis

HEMPEL, E. 1953. *Nikolaus von Kues in seinen Beziehungen zur bildenden Kunst*. Berlin: Publisher.

LEONARDO DA VINCI. 1907. Thoughts on Art and Life. *The Burlington Magazine for Connoisseurs* **11**, 49 (April 1907).

———. 1956. *Treatise on painting*. Trans. A. P. McMahon. Princeton: Princeton University Press.

LINDBERG, David C. 1976. *Theories of Vision, from Al-Kindi to Kepler*. Chicago and London: The University of Chicago Press.

LOOS, Adolf. 1998. *Ornament and Crime: Selected Essays*. Adolf Opel, ed. Riverside CA: Ariadne Press.

LYNN, Greg. 1998. *Folds, Bodies & Blobs, Collected Essays*. Belgium: La letter vole.

MARTINI, Francesco di Giorgio. 1480-82? *Trattato di architettura civile e militare*. Codex Magliabechiano II.I.141, Biblioteca Nazionale di Firenze.

REISER + UMEMOTO STUDIO. 1998. Tokyo Bay Experiment, Columbia GSAP, New York.

RICHTER J. P. and I. A. RICHTER. 1939. *The Literary Works of Leonardo da Vinci*. London.

STEADMAN, Philip. 1979. *The Evolution of Designs: Biological analogy in architecture and the applied arts*. Cambridge: Cambridge University Press

THOMPSON, D'Arcy. 1993. *On Growth and Form*, J.T. Bonner, ed. Cambridge: Cambridge University Press.

VITRUVIUS, Marcus Pollio. 1960. *Ten Books on Architecture*. New York: Dover Publications Inc.

WITTKOWER, Rudolf. 1988. *Architectural Principles in the Age of Humanism*. New York: St. Martin's Press.

ZELLNER, Peter. 1999. *Hybrid Space. New Forms in Digital Architecture*. London: Thames & Hudson.

About the author

Vesna Petresin Robert studied architecture and music and gained her PhD in temporal aspects of architectural composition. She has lectured in architecture, design and visual theory, most recently at the UCL (The Bartlett School of Architecture) and the University of Arts London (Central Saint Martins School of Art and Design). With Laurent-Paul Robert, she is co-founder and director of Rubedo, a London-based trans-disciplinary platform for research and creative practice. Their current research focuses on visual and design theories, creativity methods, geometry, digital technologies, with R&D projects ranging from bionic design and environmental engineering to parametric modelling and immersive environments. Rubedo are consultants to Ove Arup and Partners and Double Negative Visual Effects. Their creative practice includes art, film, design, sound and performance and has been published and shown internationally, recently at the Cannes Film Festival and the Beijing Architecture Biennial.

Christopher Glass

38 Chestnut Street
Camden, ME
04843-2210 USA
chris.glass@verizon.net

Keywords: Leonardo da Vinci,
Buckminster Fuller, Kenneth
Snelson, Rafael Guastavino,
lattices, tensegrity, vaulting, cast
iron, octet truss

Research

Leonardo's Successors

Abstract. Ideas similar to Leonardo's for lattice structures can found many later practical applications (Buckminster Fuller's domes, the Zome geometry of Steve Baer from the Whole Earth days, the Tensegrity structures based on the sculpture of Kenneth Snelson, as well as the Catalan vaulting traditions of Gaudi and the Guastavinos.

Introduction

Leonardo's domed wooden roofs are a product of the intense energy with which Leonardo examined the world around him and looked for ways to exploit basic principles for mechanical advantage. He was very conscious of the examples of the past, but even more excited by stimuli from natural organisms. The system he developed for the domes is at the same time a critique of past efforts to create roofed spaces without columns, and a precursor of systems it would take centuries for later inventors to rediscover. The essence of these drawings is the attempt to span relatively large open spaces with simple repeatable elements that do not require much labor to make or to assemble. What makes his system elegant and "modern" is that the idea derives from the construction sequence and the underlying geometry, and does not depend on sophisticated construction techniques or expensive materials.

Leonardo in Florence was inescapably aware of Filippo Brunelleschi's achievement in creating the dome of the Duomo. It was the wonder of the age and the emblem of the new thinking we now call the Renaissance. Brunelleschi's machinery for building the dome had as much influence on Leonardo's thinking as the achievement of the dome itself did. For an ambitious designer in Florence there would be no more such vast commissions, but the role of all-around problem solver was one the Florentines respected and one for which Leonardo was well suited, with his wide-ranging interests and uncommon ability to make connections between the working principles of organic and inorganic systems. Rivers, humans, birds, bridges, buildings, were all subjected to his analytical eye and his irresistible urge to tinker. If in many cases these analyses never went beyond the sketchbooks of the codices, the mental habits displayed there were in play everywhere he was asked to go.

The genius of Brunelleschi's dome was that it had solved the problem of keeping a large masonry dome from collapsing by a completely new method. As they are being built, domes want to fall inwards, and when they are complete they want to explode out at the base. The new system used stone and timber tension chains buried in the rings of the dome to resist the outward bursting pressure, and the successive layers of the dome were built as horizontal circular arches which resisted the tendency of the masonry to fall inward while the structure was incomplete. It was a dramatic balancing act.

The Romans had thrown mass at the problem, using formwork and fill to support concrete and brick shells. Hadrian's engineers made the dome of the Pantheon thinner as it went higher, had used square coffers to stiffen the shell, and even used hollow jars at the top to lighten the load. Even so, the perimeter at the base started to show signs of cracking, so the engineers added the outer rings that give the Pantheon its characteristic profile, in order to overload the base and literally overpower the outward thrust. It was a solution appropriate to the mindset of empire. It used the abundance of cheap labor produced by the imperial system to compensate for an incomplete understanding of how structures work.

The architects of the Gothic cathedrals had developed a more sophisticated idea of how to counterbalance loads with other loads, and how to use ribs to support thin shells of stone blocks. The ribs allowed the formwork to be much lighter, but the system required that the ribs be locked in place by the central bosses before the scaffolding could be removed. The machinery for hoisting the stones to the height of the work area was not much more advanced than that of the Romans, so the size of the blocks tended to be small, and the whole construction depended on balanced compression carried from boss to base. Irwin Panofsky's brilliant essay *Gothic Architecture and Scholasticism* details how the articulation of Gothic structure is analogous to the scholastic subdivision of syllogistic explication of the universe as a creation and emanation of the mind of God [Panofsky 1957: 34-35, 58-60].

The challenge of the Florentine dome was that it did not have a way to brace the exterior against the outward-pushing bursting pressure the huge vault would place on the drum, which had already been built. Further, the drum was so high and so wide that filling it with scaffolding or earth as the Romans would have, or with a timber frame supported on the drum as was Gothic practice, were both beyond the resources and the technical ability of the builders. Scaffolding would collapse under its own weight, fill would burst the walls, and timbers to span the space couldn't be set in place (fig. 1).

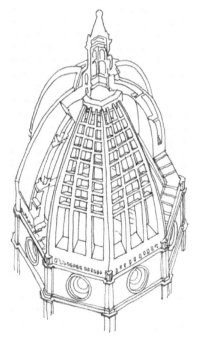

Fig. 1. Diagram of dome structure. All illustrations are by the author

Brunelleschi solved the problem with horizontal rings that could be built sequentially and support themselves. He also devised machines that could continuously raise not only the bricks and mortar but the long stones he needed to lock together to create tension "chains" around the compression rings. His design brought together a new understanding of curved structures, derived from study of the Ptolemy atlas of the spherical world, and the ability to invent mechanisms to solve problems of transmitting mechanical force which came from his experience as a metalworker. Both what to build and how to build it were his ideas and they changed the world.[1]

The problem is that all this ingenuity still took a lifetime and large amounts of material and capital. It was not suitable for daily use in marketplaces and workshops. Leonardo's idea, on the other hand, would work immediately, simply, and even demountably. Though the model he proposed wasn't as big as the Duomo (27 meters as opposed to the Duomo's 43.7 meters and the Pantheon's 43.3 meters), the system did not produce bursting stresses and could presumably have been made as large as needed.

Unfortunately, it didn't catch on. There are references to a portable bridge for military use that he designed using a similar construction technique, and there is also another intriguing sketch that shows a structure composed of straight elements held in position by some kind of cable, whether as an arched bridge or a curved roof is hard to tell (fig. 2). This is especially suggestive for later tensegrity structures, because it appears to have the cables in tension supporting beams in compression, but it's hard to tell exactly what is going on in these figures. It's another Leonardo mystery.

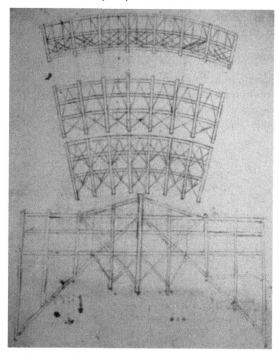

Fig. 2. Leonardo da Vinci, Ms. B of the Insitut de France, f. 29 v

As far as any related experiments with this kind of reciprocal structure, in which beams appear to support each other, there isn't much. A sketch on fol. 23r of Villard de Honnecourt's invaluable notebooks shows a roof structure which uses the "seed" of Leonardo's right-angled pattern as a way of using beams to support each other around the open well of a courtyard (fig. 3).

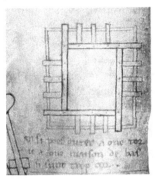

Fig. 3. Villard de Honnecourt, fol. 23r (detail)

This pattern, I am told, also appears in the music room of the Palazzo Piccolmini in Pienza, built by Bernardo Rossellini, probably between 1458 and 1464.[2] In both these cases, though, it is merely the seed. Leonardo's invention was to discover that the basic four-beam structure could be replicated by mirroring and offsetting to create a structure of essentially unlimited extension. But apart from the sketch, there is no evidence that Leonardo ever built one of his structures, and certainly his idea was not adopted by others.

Wren's workarounds

So what other solutions were there? Primarily there were timber trusses, a more polished version of traditional timber framing in which diagonal braces were combined with complex joint details to created frames that would span space. These were dependent on good quality wooden beams, and trees were grown especially for timber frameworks.

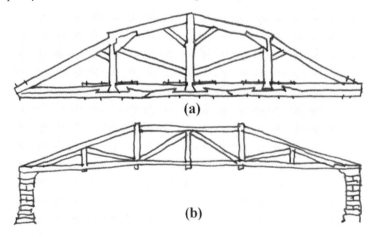

Fig. 4. a) Diagram of truss by Wren; b) Diagram of truss by Palladio

In 1669 the young Christopher Wren adapted a system developed by John Wallis for a "geometrical flat floor" to create the truss for the 21.3 meter clear span of the Sheldonian Theater at Oxford. According to his contemporary, Robert Plot [1677], it was "perhaps not to be parallel'd in the World" [Tinniswood 2001: 104] and considered a technological marvel of the same kind as the Florence dome (fig. 4a). In fact, the technological innovation was simply the splicing together of shorter beams using variations on "scarf" and dovetail joints, together with iron bolts to hold the joints together. This system may have been new to England, but Leonardo had sketched something similar in the *Codex Atlanticus* (344 verso a), and scarf joints had been used in the ceiling of the Doge's Palace in Venice at least by 1424 [Mehn 2003].

The roof itself was braced rather than genuinely triangulated, as was for example the bridge truss in Andrea Palladio's books. Palladio drew the bridge of Cismone [Palladio 1738, Bk. III, ch. VII, pl. III] (fig. 4b), though he stops short of claiming it as his own design, and accurately described the action of the truss members as working reciprocally ("... those are also supported by the arms that go from one colonello to the others, whereby all the parts are supported the one by the other; and their nature is such, that the greater the weight upon the bridge, so much the more they bind together, and increase the strength of the work...." [Palladio 1965: 65]). Wren's upper framing, however, was not a true truss because it did not use the diagonal rafters as part of the structural bracing.[3]

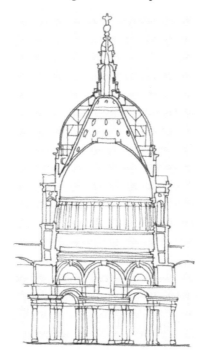

Fig. 5. St. Paul's Cathedral, London

When it came Wren's time to design a dome on the scale of the Cathedral of Florence, he used what we would call a "workaround" to address the problem of bursting. Instead of building a circular dome, he set a brick cone on a base chain (fig. 5). The stresses in a cone are transmitted directly along the length of the cone to the base, so it did not have to be tied as it went up. A shallow masonry shell formed the interior dome, and a copper skin over a timber framework formed the outer dome. So Wren's structures, while innovative and clever, evaded the question of how to span large areas simply.

Cast iron

The real breakthrough to a system with the elegance of Leonardo's simple beams came in the village of Coalbrookdale, where in 1759 Abraham Darby, John Wilson, and T. F. Pritchard used repeated cast iron components to span more than 30 meters (fig. 6). The new material and the idea of prefabricating replaceable elements led to an explosion of new structural ideas for glasshouses and exhibition halls.

Fig. 6. Coalbrookdale Bridge

By the middle of the nineteenth century, the ideas generated by the Coalbrookdale bridge would culminate in Joseph Paxton's Crystal Palace of 1851. Paxton, a designer of glasshouses, is reported to have designed the hall in only ten days, using techniques he had already developed. Its modular construction covered 770,000 square feet of space and made use of shallow iron trusses. The diagonals of timber trusses, like those of Palladio's bridges, were added to horizontal and vertical members to create a very lightweight but strong web-like beam that stood in for the solid beams which casting techniques could not produce. Prefabricated sections could be bolted together in place, and a system of trolleys on rails enabled the roofers to install the glass panels with a minimum of effort (fig. 7). After the exhibition the palace was disassembled and re-erected at Sydenham Hill in South London, where it stood until destroyed by fire in 1936.

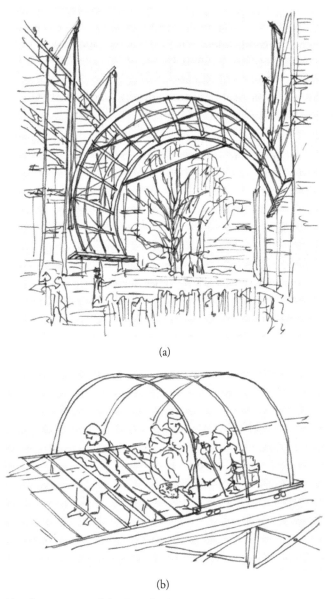

(a)

(b)

Fig. 7. Assembly of components of the Crystal Palace. a) Raising the arches; b) installation of the glazing

The Crystal Palace, in its simple elements easily assembled and disassembled, is the direct heir to Leonardo's timber grid. The system it embodied would become the standard for construction of large areas like railroad stations and exhibition halls well into the twentieth century, and its more humble variant of the open-web joist would be the material of choice for inexpensive market buildings and offices – just the kinds of buildings Leonardo had intended for his wooden domes.

The more general idea of interchangeable cast iron components would be adapted to more conventional buildings as well. In the 1850s in New York James Bogardus developed a system for commercial construction, using designs that appeared to be classical carved stone. In an engraving from 1856 he illustrated the strength and flexibility of the system by showing a façade with half its pieces missing, but which could still support itself.

After Bogardus, no longer would structural integrity depend on stacking masonry pieces and relying on the geometry of arches and lintels to hold them in place. Bolts could be used to suspend elements in tension, as well as to stabilize them in traditional compression structures. It would take a few years before the implications of the new freedom would begin to dawn on designers, but in the meanwhile cast iron became a means of cheaply imitating carved stone masonry, while providing strength and durability far beyond the capacity of masonry alone.

This idea of using a concealed or disguised iron structure to support buildings that appear to be traditional masonry buildings led to the early skyscrapers of Chicago and New York, but it was used even earlier in Thomas U. Walter's design for the enlarged dome of the U.S. Capitol, built during the Civil War. A section through Walter's dome shows that the system is a variation on Wren's St. Paul's (fig. 8). The structural skeleton is a nearly conical array of trusses, below which is an inner dome with coffers cast to resemble the stone coffers of the Pantheon, and above which are braces supporting an outer skin of cast iron resembling Wren's copper dome.

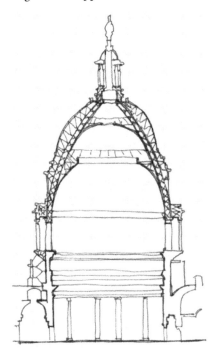 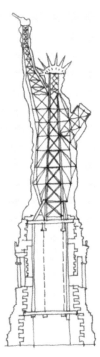

Fig. 8. Dome of the U.S. Capitol   Fig. 9. The Statue of Liberty

Even Frederic Auguste Bartholdi's Statue of Liberty (conceived in the 1870s but not completed until 1886), which seems to be a huge version of a cast bronze figure, is a thin copper skin, attached with clever clips that prevent electrolysis between iron and copper to an iron frame designed by Gustave Eiffel (fig. 9). A large part of the fame of Eiffel's tower built in Paris in 1900 is a result of his letting the structure speak for itself rather than using his engineering skill to disguise an iron frame within a conventional envelope. Cast iron began to break free of its imitative role.

The most dramatic application of these techniques was the suspension bridge. Thomas Telford had pioneered the form, and John Roebling used it to build the Brooklyn Bridge, completed in 1883, and several others, establishing the type in America. Before emigrating to America, Roebling had studied with Friedrich Hegel; I have always seen his suspension bridges as the physical embodiment of Hegel's idea of the dialectic struggle in which a thesis is opposed by an antithesis, producing a new synthesis. In the suspension bridge the vertical tower in compression supports the cables in tension, which in turn support the bridge deck, which would be impossible without the other supporting elements. The towers are expressed as Gothic survivors of an earlier age, while the cables are unapologetically unadorned. Thus the structure spans the ages as well as the spectrum from extreme compression to extreme tension. This conceptual separation of tension and compression would be the key to a new understanding of structural form at the end of the next century.

Fuller's domes

Buckminster Fuller's geodesic domes are variations on the triangulated rigid metal framework. Though Fuller promoted himself as an innovator in the league of Leonardo and Brunelleschi, his system was fundamentally the application of the idea of triangulation to spherical structures. His domes take the geometry of the truncated icosahedron, a form familiar as the soccer ball, and subdivide the hexagons and pentagons into irregular triangles which can then be made more rigid by converting each triangle into a shallow tetrahedron. While the result appears novel, the principle of the frame made rigid by diagonal bracing has been the fundamental engineering principle of design since Palladio's bridge.

Before Fuller developed his tetrahedral system, telephone inventor Alexander Graham Bell had spent his later years investigating the possibilities of vast tetrahedral networks. Unfortunately for Bell, his vision was of using the structures as vast aerial kites for transporting cargo, an idea dependent on either prevailing winds or an as-yet undeveloped motor. The Wright Brothers' warped wings (fulfilling another idea prefigured in Leonardo's works) would spell the end of the tetrahedral kite. The tetrahedral grid would, however, prove to be one of the major structural innovations in the twentieth century.

Fuller's obsession with spherical domes became a profound limitation to the spread of his system to the world outside theme parks and world's fairs. A few circular halls, such as the 1957 Kaiser Dome in Honolulu, were built, but the major application of Fuller's system became enclosures of sewage treatment tanks and the proliferation of small dome houses among proponents of the counterculture of the 1960s and later.

One attempt to break out of the sphere was the use of the Zonohedral geometry by Steve Baer in New Mexico and Colorado in the 1960s [Kahn 1972: 102]. What he called "Zomes" are polyhedra with a complete circumferential zone of edges that are parallel to each other (fig. 10).

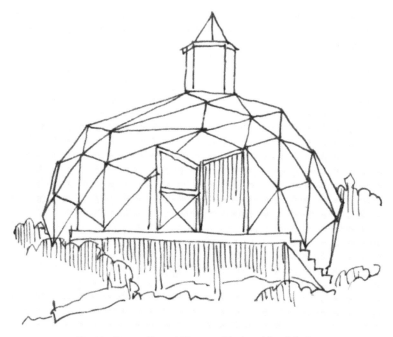

Fig. 10. Garnet Crystal Zome at Placitas, New Mexico

Baer realized that such domes could be stretched out of shape by elongating or shortening the parallel edges, and that domes could be joined into clusters using the parallel zones as links. The rhombic triacontahedron was the shape he found most suitable. While this generated some flexibility, it was not enough to make the dome a popular alternative to the rectangular box, either for homes or for convention halls. Remembering the name of the shape was almost as difficult as remembering the proportions of the struts.

The domes remain a vehicle for unconventional expression, outside the mainstream of construction technology. In many ways, Fuller's own writings and polemical stances helped to ensure they would remain there.

The octet truss

One system Fuller christened the "octet truss" did become a widely used structure, precisely because it was adaptable to rectangular and irregular spaces. As with the geodesic dome, the truss was a variant of the triangulated beam, with the diagonals spanning from beam to beam to create square-based pyramids that Fuller perceived as octahedrons cut in half (fig. 11).

Though Alexander Graham Bell had done something similar with tetrahedra, and Louis Kahn would use a tetrahedral concrete truss in his Yale Art Gallery, the octet form superseded the tetrahedron because its rectilinear geometry of staggered squares was more adaptable to the usual rectangles of modern floor plans. The octet would be refined by numerous manufacturers for use as roofing systems and display structures. Biosphere II is a good example of this kind of structure. It combines straight areas and curved sections, all based on the octahedral/tetrahedral geometry of the rigid truss.[4]

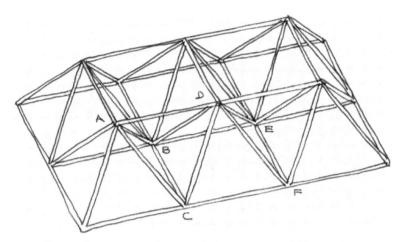

Fig. 11. Octet truss. ABCD = tetrahedron; BCDEF = half octahedron

One of the more flamboyant uses of the octet truss is Philip Johnson's Crystal Cathedral, built for evangelist Robert Schuller in Pasadena in 1980. The name is a clear reference to the Crystal Palace, and the space has the same quality of expansive transparency. It is emphatically not a dome, but a prismatic irregular structure of rectilinear elements, so it achieves the goals implicit in Leonardo's grid sketches: simplicity, flexibility, ease of construction, even, should it be necessary, ease of deconstruction.

One aspect of the Crystal Cathedral is that a whole section of wall had to be able to be opened to the parking lot, so people parked in their cars could see the pulpit. Johnson's office contacted NASA to find out how the Cape Canaveral Assembly building doors worked, and NASA engineers told them how to make the basic mechanism, but the doors themselves are sections of the same rigid octet truss.

Concrete

All of these systems were based on steel struts with various skins, usually glass or sheet metal. The other material of the twentieth century, reinforced concrete, was also used to span great distances, but the labor to build the formwork and to place the wet concrete made the material less attractive than metal.

One of the greatest concrete domes is also one of the earliest, Max Berg's Centenary Hall in Breslau of 1912-13 (fig. 12). Robert Hughes [1980] tells the story that when the formwork was to come off, the workers refused, fearing the dome's collapse, and Berg himself had to remove the first props before the workers would continue. The shell, with its ribs and concentric rings, is the skeleton of Brunelleschi's dome. Reinforced concrete uses embedded steel to resist the bursting and bending stress that masonry is so bad at handling. The concept of using concrete in compression and steel in tension marked a step on the way to thinking about those two forces in different ways, which would free engineering from rigid structural concepts. Brunelleschi had understood the function of the "chains" of stone that bound his dome, but had used hard stone with secretly conceived joints. Tie rods and iron chains had been used for centuries, but the innovation of embedding the thin rods in the concrete freed the engineer to create what were in effect long "stone" beams and shells.

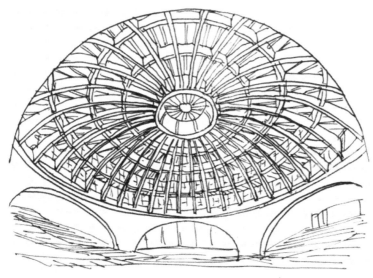

Fig. 12. Centenary Hall, Breslau

The poet of concrete was of course Pier Luigi Nervi, whose graceful structures allowed the mass of concrete to float almost effortlessly over vast spaces, and he pioneered the use of precast elements which made construction less difficult. Nervi's structural ideas were often based on the lamella structure of interlaced continuous beams. While not specifically a triangulated structure, the lamella dome could have its stresses calculated using techniques that did not deviate from standard practice.[5] Today the prestressed and precast tee is widely used, though usually for parking garages, and precast concrete is more widely used as a surfacing material than a structural one.

The bóveda tabicada

Apart from steel frameworks and the occasional concrete ribbed structure, there was one other system prevalent in the late nineteenth and early twentieth centuries that fits the description of Leonardo's lattice: the *tabicada* or tiled dome. *Bóvedas* were traditional masonry domes derived from vernacular Arabic construction and used in Spain for such structures as wine cellars. Carried to Mexico by Spanish masons, they were used occasionally for house roofs. The technique allows a mason to form a domed roof without extensive formwork. Using quick-setting mortar and lightweight bricks, he can place one brick at a time in space, waiting long enough for the mortar to grip before moving on to the next brick.

In Cataluña the system was refined by the substitution of flat clay tiles for bricks, permitting very thin shells to be built over relatively large areas. The technique was used by Antoní Gaudí in several buildings, most spectacularly in his school building on the grounds of the Sagrada Familia church in Barcelona. Its undulating bóveda shell is supported by a central girder and straight rafters that form the frame for the shell. Gaudí never seems to have allowed the shells to become the whole structure, however. He depends on ribs to support the shells, as in the roof structure for the attic of Casa Milá and the crypt of the

Guëll chapel. The ribs themselves are built of the same tile, but used as straight compression membranes.

Le Corbusier noticed and sketched Gaudí's school roof, and then adopted the tabicada vault for his Maison Jaoul of 1955-57. We think of Le Corbusier as using reinforced concrete, but here he used this masonry construction technique in one of his important late works.

Guastavino vaulting

The man who brought the bóveda tabicada into the architectural mainstream was Rafael Guastavino Moreno, a Catalan of Genoan ancestry who began by building fire- and damp-proof vaults for wineries around Barcelona before emigrating to the United States in 1881. There he promoted the technique as a means of fireproofing steel frame construction, but he soon developed a complete structural system. He was able to convince McKim Meade and White to use his vaults in the Boston Public Library of 1895, and soon he and his son (also named Rafael) were supplying domes and vaults for many of the most important buildings in the United States.

Among the many projects to use what came to be called Guastavino vaulting were the Cathedral of St. John the Divine in New York by Heins & LaFarge, the Christian Science mother church in Boston, and the Shrine of the Immaculate Conception in Washington, while at the same time the tiles lined subways and train stations [Huerta 1999] and indoor swimming pools.

Guastavino achieved the geometric regularity not typical of traditional bóvedas by using a lightweight system of ribs and spacers. Unlike formwork for concrete, the frame did not support the weight of the shell but merely provided a geometric reference for the masons. At St. John a stiff wire was fixed to a weighted plate suspended at the radius point of the spherical dome and used to check the radius of the dome at each tile (fig. 13).

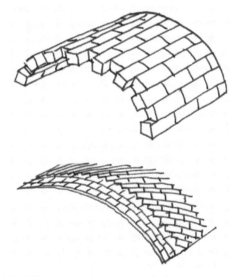

Fig. 13. a) Conventional masonry vault; b) Bóveda tabicada

Guastavino dealt with building codes by staging load tests. The system proved capable of supporting loads far in excess of structural needs, while being flexible enough to build hemispherical and shallow domes and curved planes such as the helix of a curved stair. To satisfy the code officers, Guastavino developed graphical analyses of the stresses of the dome based on conventional engineering.[6]

Guastavino vaults were even used by McKim Meade and White to restore Thomas Jefferson's Rotunda at the University of Virginia after it burned in 1895. They were used to fireproof the floors and porch roof as well. John Russell Pope, the original architect of Jefferson's memorial in Washington, used the system in Washington for the Masonic Hall, a pyramidal structure based on the Mausoleum of Halicarnassus. An unlikely candidate for a dome system, the building was highlighted in an advertisement for the Guastavino Company as being similar in its double-layered construction to, of all things, Brunelleschi's dome.

Also in its advertising, Guastavino Company took on its main competitor, steel framing. In a graphically compelling side-by-side section drawing, the ad says that the system is "simple, economical, and the necessary materials can always be delivered promptly" – the last because they did not have to be fabricated specially for the project.

The Guastavinos were not the only ones to use tabicada techniques in modern times. In Spain Luis Moya built several buildings using vaults with and without tile ribs. For the church of Santa Maria de la Iglesia of 1966-69, he developed an elegant mechanism using a rotating steel frame to align the tiles. In Havana in 1961, the Cuban architect Ricardo Porro began the elaborate complex of the National Schools of Art, which linked domes of several sizes with a sinuous set of corridors roofed by tiled tunnel vaulting (fig. 14).

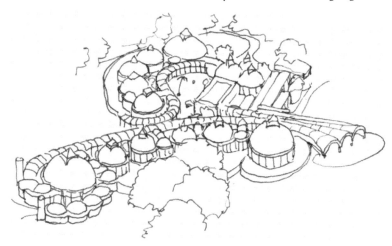

Fig. 14. Plan of Porro's project for the National Schools of Art

The elder Rafael Guastavino, having worked on vaults for Richard Morris Hunt's Biltmore, the Vanderbilt summer chateau near Asheville, North Carolina, had built a retirement home and studio in nearby Black Mountain. He worked with Hunt's local architect, Richard Sharpe, to build the church of St Lawrence, which features a large elliptical tile dome and several smaller chapels and helical stairways. When he died, he was buried in a tiled tomb in the church.

Snelson's tensegrity

By a coincidence of history, in 1949, the young Oregon sculptor Kenneth Snelson attended a summer workshop at the Black Mountain School, which by then was the home of several refugee Bauhaus figures, notably Joseph and Anni Albers. The architect scheduled to teach was replaced at the last minute by Buckminster Fuller. Snelson showed him a sculpture he had been working on using wooden struts connected by cables. Fuller asked to keep it, and shortly was touting what he called "tensegrity" geometry, which he privately told Snelson had been Snelson's idea, but publicly refrained from attributing to anyone but himself.[7]

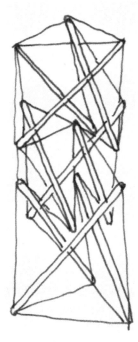

Fig. 15. Snelson patent drawing

Snelson in 1960 patented the system (fig. 15), which he more accurately if less memorably called "continuous tension, discontinuous compression structures". He clearly spelled out in his patent and in his sculptural works over the next half century his understanding of the significance of thinking separately about tension forces and compression forces in designing structures. He has made the analogy that the body should be considered as having a compression structure of bones linked by a tension structure of tendons and muscles. Structural freedom can be achieved by conceptually separating the two forces. This was the insight that had led to the suspension bridge, but Snelson's explicit understanding of it made much more flexible structures possible.

Snelson has described his system as based on weaving techniques, where the connections between members are determined by the ways in which they overlap or interweave.[8] Analysis of woven structures allowed him to think about polyhedral analogies, with edges of polyhedra conceived as fibers that bypassed each other in regular ways. And separating compression from tension allowed him to convert what he called "weave polyhedra" to tensegrity polyhedra using compression struts connected by cables. Modules could be interconnected by stacking and extending. The interlaced framework of his structures bears a remarkable formal similarity to Leonardo's grids, especially in the way that the beams must overlap in a specific sequence in order to work. Like Snelson's sculptures, the frames can be right- or left-handed, depending on the way the beams overlap.

So, from analysis of the most widespread structures man has made – weavings – Snelson has developed a theoretical system capable of using, as Leonardo had wanted, simple elements easily connected to produce structures of great flexibility and variety.

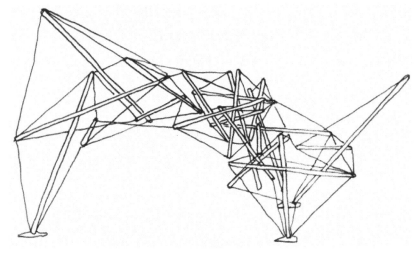

Fig. 16. Snelson's *Free Ride Home*

Snelson's most well known sculpture is the *Needle Tower* of 1968 at the Hirshhorn Museum in Washington. A more exciting example is the *Free Ride Home*, one of several at the Storm King Sculpture Park in New York. While the needle tower is dramatic, *Free Ride Home* (fig. 16) shows the possibilities for irregular shapes the system allows.

Snelson insists that the true utility of the tensegrity system is for dramatic sculpture forms of the kinds he creates. More sober engineers, however, have used his system to span the large spaces like those of athletic fields – the same use that Fuller envisioned for his domes. Some twenty years after the steel lamella dome of the Astrodome, David Geiger designed stadiums for the Seoul Olympics. The Fencing Arena in particular shows the basic tensegrity system: a compression ring at the top of the stands supports cables that hold the tops and bottoms of vertical compression struts suspended over the arena (fig. 17). From the tops of the struts another similar system of cables and struts extends further into the space. Yet another set extends further in, until the system converges at a central hub. The dome is given its final shape by tightening the bottom cables in sequence, as shown in the figure.

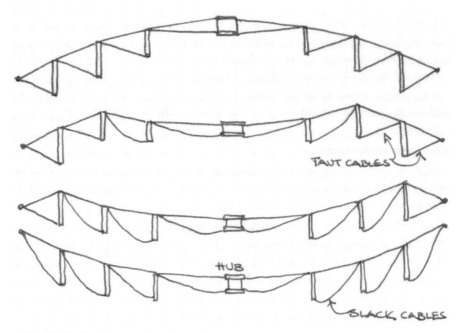

Fig. 17. Fencing Arena section. Circumferential cables connecting bases of masts not shown

This dome and the several others built by Geiger and by Weidlinger Associates take Snelson's poetic spatial constructions and turn them into economical utilitarian roofing systems, competitive with inflatable or cable-hung fabric structures. Cable-hung structures are a development of the suspension bridge, with compressions masts and tensions cables used to support a roof rather than a road. The Millennium Dome (now the O2) is a recent example of that system.

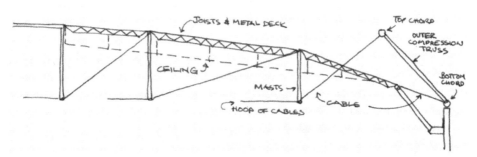

Fig. 18. Georgia Dome

The tensegrity system is not limited to flexible fabric roofs but can accommodate conventional roofs made of panels supported by the simple open web joists and corrugated steel roofing of factories and warehouses. The domes need not be circular. Weidlinger's Georgia Dome is an oval, 235 by 186 meters (fig. 18).

As a new century begins we have extraordinary capacity to invent new structural shapes using existing understandings of compression and tension — Snelson's bones and sinews.

While the current, rather conventional uses of tensegrity domes are exciting by virtue of their lightness and immense scale, if we look at *Free Ride Home* and think of some of the formal adventures of people like Frank Gehry, Zaha Hadid, and Santiago Calatrava, the possibilities of incorporating tensegrity structural techniques with architecturally adventurous forms would excite even Leonardo.

And perhaps, especially in parts of the world where labor is more available than manufactured materials, the Guastavino dome and even the Leonardo grid might make a comeback.

The Leonardo Sticks Project

After attending the conference on Rinus Roelof's rediscovery of Leonardo's domes I returned home full of enthusiasm for the system. I made myself a set of Rinus's small sticks and showed them off as often as I could find occasion.

One person I showed them to was an architectural client of mine, Joseph Stanislaw, who became as excited as I was about them. He in turn had a friend who had a company reproducing classic toys and games. Joe and I decided to use his connections to have sets of the sticks manufactured in China. We set up a small family company to handle the legal and logistical work, and I designed the box and information for the set. We offered royalties to Rinus on the sales of the sets, which of course we envisioned would take off as the latest craze.

Unfortunately for our enterprise, neither Joe nor I had the time to devote to marketing the sticks effectively, and despite several promising possibilities we have had few actual sales, either directly or to wholesale buyers. After four years, we have decided to liquidate the company, with several hundred sets from our original order still unsold.

Like Rinus, I have been demonstrating the sticks in various venues, notably the classes I have taught at Bowdoin College. Everywhere they are demonstrated the attract attention and interest. One reaction that has been of special interest is the idea that the system should be adapted to emergency shelters. Especially in climates where bamboo is available, a sizeable shelter could be quickly put together from available materials.

I think there is a place for a professionally marketed sticks kit, and an opportunity to develop an emergency shelter system. What would be most useful for Leonardo's system to enter the public consciousness, however, would be a large structure based on the system. What stand in the way of that is what hampered Fuller and Guastavino: an accepted means for calculating the stresses and therefore assuring the stability of the structure. We have seen that it works. Now the task is to prove it.

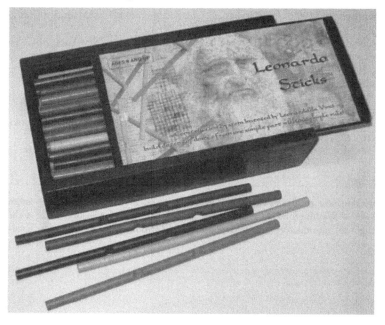

Notes

1. [King 2000] provides a good introduction to the splendid adventure of the Duomo.
2. After I lectured on this material at the Bath Scientific and Literary Institute in October 2007, Nicholas Lewis told me about the Piccolomini. I have not had an opportunity to verify whether this is in fact a reciprocal structure or a decorative ceiling, but given its date it is not inconceivable that Leonardo might have seen it.
3. While on the subject of Palladio's bridges, I would note the similarity between his arched bridge, plate V of Book III, described in chapter VIII, which bears a remarkably similarity to the Leonardo sketch described above.
4. In an interesting reversal, Biosphere's successor the Eden Project in Cornwall, whose enclosure is by Nicholas Grimshaw, uses a newer flexible version of the geodesic dome. The flexibility derives from separating the regular polygons of the skins from the bracing system. This system has similarities to the tensegrity systems discussed later in this article.
5. For this reason the first major sports arena in America, the Astrodome in Houston, would use a lamella dome rather than a geodesic dome. For a discussion of lamella structures and the Astrodome in particular by L. Bass, see [Davies 1967], available online at: http://www.columbia.edu/cu/gsapp/BT/DOMES/HOUSTON/h-lamel.html.

6. Gaudí had used similar techniques to determine the slope of his retaining wall at the Parque Guëll, and in general to guide his departures from rectilinear geometries. See [Sweeney and Sert 1960: 74].

7. This information is from a letter from Kenneth Snelson to R. Motro, published in *International Journal of Space Structures* (November 1990). It is available at *http://www.grunch.net/snelson/rmoto.html* .

8. See http://www.kennethsnelson.net/main/structure.htm for his description of the principles involved.

References

DAVIES, R.M., ed. 1967. *Space Structures.* New York: John Wiley & Sons.

HUERTA, Santiago, ed. 1999. *Las Bóvedas de Guastavino en America.* Madrid: Instituto Juan de Herrera.

HUGHES, Robert. 1980. "Trouble in Utopia", *Shock of the New* series. New York: BBC/ Time Life.

KAHN, Lloyd, ed. 1972. *Domebook 2.* Bolinas, CA: Shelter Publications.

KING, Ross. 2000. *Brunelleschi's Dome.* London: Chatto &Windus Random House.

MEHN, Daniel J. 2003. Letter. *Old House Journal* (August 2003): 12.

PALLADIO, Andrea. 1965. *Four Books of Architecture* (1738, Ware trans.). Rpt. New York: Dover Publications.

PANOFSKY, Irwin. 1957. *Gothic Architecture and Scholasticism.* New York: Meridian Books.

PLOT, Robert. 1677. *Natural History of Oxfordshire.*

SWEENEY, James Johnson and Josep Lluís SERT. 1960. *Antoni Gaudí.* New York: Praeger.

About the author

Christopher Glass is a architect with a one-person practice in coastal Maine. He attended Saint Alban's School in Washington, D.C., studied philosophy at Haverford College and architecture at Yale. He taught an introductory architecture studio at Bowdoin College, from which he retired in 2005.

Rachel Fletcher

113 Division St.
Great Barrington, MA
01230 USA
rfletch@bcn.net

Geometer's Angle

Dynamic Root Rectangles Part Two: The Root-Two Rectangle and Design Applications

Keywords: descriptive
geometry, diagonal,
dynamic symmetry,
incommensurate values,
root rectangles

Abstract. "Dynamic symmetry" is the name given by Jay Hambidge for the proportioning principle that appears in "root rectangles" where a single incommensurable ratio persists through endless spatial divisions. In Part One of a continuing series [Fletcher 2007], we explored the relative characteristics of root-two, -three, -four, and -five systems of proportion and became familiar with diagonals, reciprocals, complementary areas, and other components. In Part Two we consider the "application of areas" to root-two rectangles and other techniques for composing dynamic space plans.

Introduction

"Dynamic symmetry" is the term given by the early-twentieth-century artist and scholar Jay Hambidge for the system of incommensurable ratios that appear in root rectangles and that replicate through endless spatial divisions, while expressing the relationship between one level of scale and the next.[1] In contrast to passive or static symmetry, which Hambidge associates with the radial subdivisions that characterize regular geometric figures or crystals, dynamic symmetry governs the spiral-like growth exhibited in plants and shells. Hambidge observes dynamic symmetry in ancient Egyptian bas reliefs and Classical Greek pottery and temple plans.[2]

The root-two rectangle can be formed from the side and hypotenuse of a 45°-45°-90° triangle. In this article we explore the dynamic properties of this elementary quadrilateral shape and consider possibilities for design applications.

Review: Root-Rectangles in Series

When a square divides into diminishing root rectangles or when root rectangles expand from a square, elements of dynamic symmetry become apparent. Fig. 1 illustrates how expanding root rectangles develop from a square. The diagonal of the preceding square or rectangle equals the long side of the succeeding four-sided figure.[3] The short side of each root rectangle is 1. The long sides progress in the series $\sqrt{2}$, $\sqrt{3}$, $\sqrt{4}$, $\sqrt{5}$, $\sqrt{6}$... .

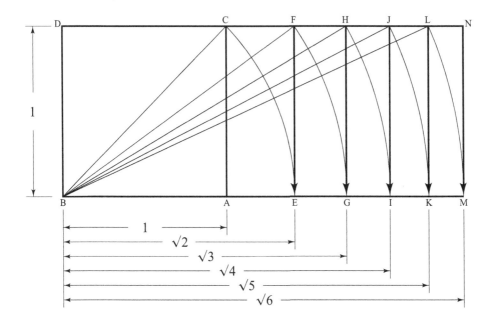

$$DB : BA :: 1 : 1$$
$$DB : BE :: 1 : \sqrt{2}$$
$$DB : BG :: 1 : \sqrt{3}$$
$$DB : BI \ :: 1 : \sqrt{4}$$
$$DB : BK :: 1 : \sqrt{5}$$
$$DB : BM :: 1 : \sqrt{6}$$

Fig. 1

Another method for obtaining root rectangles is from the radii of arcs that increase in whole number increments.[4]

- Draw a horizontal baseline AB equal in length to one unit.
- From point A, draw an indefinite line perpendicular to line AB that is slightly longer in length.
- Place the compass point at A. Draw a quarter-arc of radius AB that intersects line AB at point B and the indefinite vertical line at point C.
- Place the compass point at B. Draw a quarter-arc (or one slightly longer) of the same radius, as shown.
- Place the compass point at C. Draw a quarter-arc (or one slightly longer) of the same radius, as shown.
- Locate point D, where the two quarter-arcs (taken from points B and C) intersect.
- Place the compass point at D. Draw a quarter-arc of the same radius that intersects the indefinite vertical line at point C and line AB at point B (fig. 2).

* * *

- Connect points D, B, A, and C.

The result is a square (DBAC) of side 1 (fig. 3).

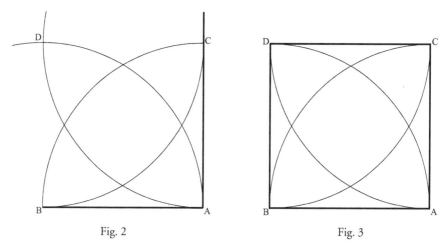

Fig. 2   Fig. 3

- Add a length of ½ unit to line DB, so that the line DM equals 1.5.
- Place the compass point at M. Draw an arc of radius MD that intersects the extension of line BA at point E.
- From point E, draw a line perpendicular to line EB that intersects the extension of line DC at point F.
- Connect points D, B, E, and F.

The result is a root-two rectangle (DBEF) with short and long sides of 1 and √2.

- Add a length of ½ unit to line DM, so that the line DN equals 2.0.
- Place the compass point at N. Draw an arc of radius ND that intersects the extension of line BE at point G.
- From point G, draw a line perpendicular to line GB that intersects the extension of line DF at point H.
- Connect points D, B, G, and H.

The result is a root-three rectangle (DBGH) with short and long sides of 1 and √3.

- Add a length of ½ unit to line DN, so that the line DO equals 2.5.
- Place the compass point at O. Draw an arc of radius OD that intersects the extension of line BG at point I.
- From point I, draw a line perpendicular to line IB that intersects the extension of line DH at point J.
- Connect points D, B, I, and J.

The result is a root-four rectangle (DBIJ) with short and long sides of 1 and √4.

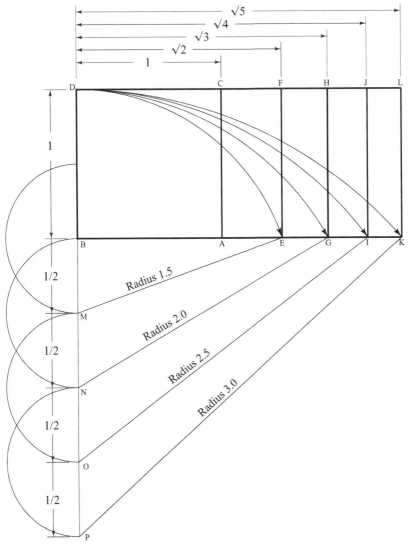

Fig. 4

- Add a length of ½ unit to line DO, so that the line DP equals 3.0.
- Place the compass point at P. Draw an arc of radius PD that intersects the extension of line BI at point K.
- From point K, draw a line perpendicular to line KB that intersects the extension of line DJ at point L.
- Connect points D, B, K, and L.

The result is a root-five rectangle (DBKL) with short and long sides of 1 and √5. The process can continue infinitely (fig. 4).

Fig. 5 illustrates how diminishing root rectangles develop from a quarter-arc that is drawn within a square.[5] The long side of each root rectangle is 1. The short sides progress in the series $1/\sqrt{2}$, $1/\sqrt{3}$, $1/\sqrt{4}$, $1/\sqrt{5}$… .

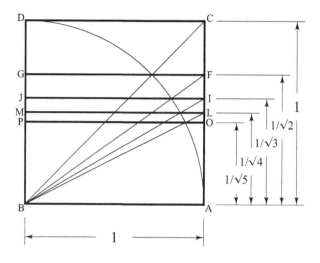

DB : BA :: 1 : 1
GB : BA :: 1 : $\sqrt{2}$
JB : BA :: 1 : $\sqrt{3}$
MB : BA :: 1 : $\sqrt{4}$
PB : BA :: 1 : $\sqrt{5}$
Fig. 5

Review: Diagonal, Reciprocal, and Complementary Patterns

At the heart of Hambidge's system is the fact that root rectangles produce reciprocals of the same proportion when the diagonal of the major rectangle and the diagonal of its reciprocal intersect at right angles.[6] The diagonals divide into vectors of equiangular distance apart, progressing in spiral-like fashion, while locating endless spatial divisions in continued proportion. The middle of any three adjacent vectors is the mean proportional or geometric mean of the other two.

A root-two rectangle divides into two reciprocals in the ratio 1 : $\sqrt{2}$. The area of each reciprocal is one-half the area of the whole (fig. 6a).

A root-three rectangle divides into three reciprocals in the ratio 1 : $\sqrt{3}$. The area of each reciprocal is one-third the area of the whole (fig. 6b).

A root-four rectangle divides into four reciprocals in the ratio 1 : $\sqrt{4}$. The area of each reciprocal is one-fourth the area of the whole (fig. 6c).

A root-five rectangle divides into five reciprocals in the ratio 1 : $\sqrt{5}$. The area of each reciprocal is one-fifth the area of the whole (fig. 6d).[7]

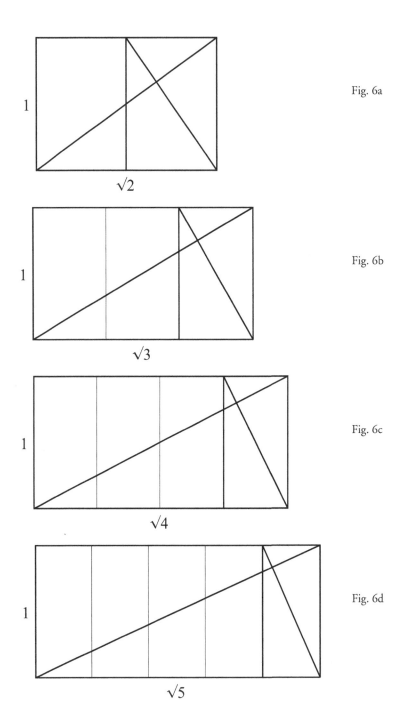

1   √2   Fig. 6a

1   √3   Fig. 6b

1   √4   Fig. 6c

1   √5   Fig. 6d

How to divide a root rectangle into reciprocals

For this demonstration we use the root-two rectangle, but the principle applies to all rectangles of dynamic proportions.

- Draw a square (DBAC) of side 1 (repeat figures 2 and 3).

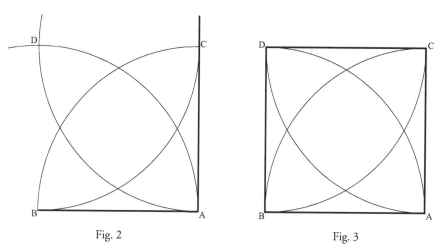

Fig. 2	Fig. 3

- Draw the diagonal BC through the square (DBAC).

The side (DB) and the diagonal (BC) are in the ratio $1 : \sqrt{2}$.

- Place the compass point at B. Draw an arc of radius BC that intersects the extension of line BA at point E (fig. 7).

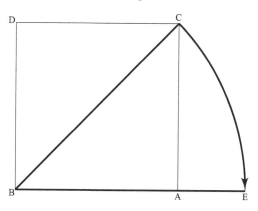

Fig. 7

- From point E, draw a line perpendicular to line EB that intersects the extension of line DC at point F.
- Connect points D, B, E, and F.

The result is a root-two rectangle (DBEF) with short and long sides of 1 and √2 (1 and 1.4142…) (fig. 8.)

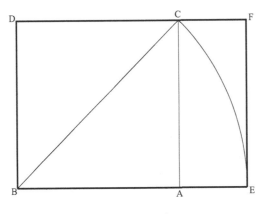

DB : BE :: 1 : √2

Fig. 8

- Locate the diagonal BF of the rectangle DBEF.
- Locate the line EF. Draw a semicircle that intersects the diagonal BF at point O, as shown.
- From point E, draw a line through point O that intersects line FD at point G.

The diagonal (BF) of the major rectangle (DBEF) and the diagonal (EG) intersect at right angles (fig. 9).

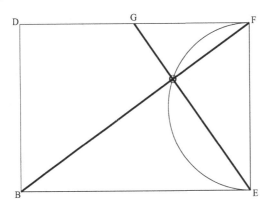

EG : BF :: 1 : √2

Fig. 9

- From point G, draw a line perpendicular to line FD that intersects line BE at point H.
- Connect points H, E, F, and G.

The result is a smaller root-two rectangle (HEFG) with short and long sides of $1/\sqrt{2}$ and 1 ($\sqrt{2}/2$: 1 or 0.7071... : 1). Rectangle HEFG is the reciprocal of the major rectangle DBEF.

The major 1 : $\sqrt{2}$ rectangle DBEF divides into two reciprocals (HEFG and BHGD) that are proportionally smaller in the ratio 1 : $\sqrt{2}$. The area of each reciprocal is one-half the area of the whole (fig. 10).

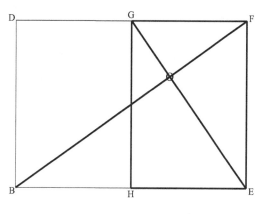

GF : FE :: FE : EB :: 1 : $\sqrt{2}$

Fig. 10

The diagonal (BF) of the major rectangle (DBEF) and the diagonal (EG) of the reciprocal (HEFG) locate endless divisions in continual proportion. A root-two rectangle of any size divides into two reciprocals in the ratio 1 : $\sqrt{2}$. If the process continues, the side lengths of successively larger rectangles form a perfect geometric progression (1 ,$\sqrt{2}$, 2, 2$\sqrt{2}$...). The side lengths of successively smaller rectangles decrease in the ratio of 1 : $1/\sqrt{2}$ toward a fixed point of origin known as the pole or eye (point O). (See figure 11.)[8]

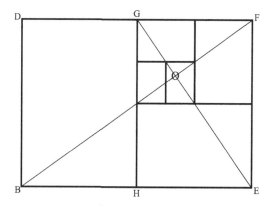

Fig. 11

- Locate the root-two rectangle (DBEF) of sides 1 and √2, and its diagonal BF.
- From point F, draw an indefinite line perpendicular to the diagonal FB.
- Extend the line BE until it intersects the indefinite line at point I, as shown.
- Connect points B, F, and I.

The result is a right triangle (BFI).

- From point I, draw a line perpendicular to line IB that intersects the extension of line DF at point J.
- Connect points E, I, J, and F.

The rectangle EIJF is the reciprocal of the root-two rectangle DBEF.

If the long side (BF) of a right triangle (BFI) equals the diagonal of a major rectangle, the short side (IF) of the triangle equals the diagonal of the reciprocal (fig. 12).[9]

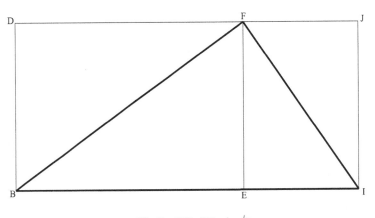

EI : IJ :: DB : BE :: 1 : √2
IF : FB :: 1 : √2

Fig. 12

Application of Areas

Definition:

Application of areas is Jay Hambidge's term for dividing rectangles into proportional components, where shapes applied on the short and long sides of a figure are equal in area.

Hambidge's technique for proportioning areas, which he traces to ancient Egyptian and Classical Greek design, produces harmonious compositions that are based on the proportions of the enclosing rectangle. An area of any rectangular shape may be "applied" on the short and long sides of a rectangle. When a square is applied on the short side, then "applied" on the long side, the new area is the same as the square, but the shape becomes rectangular. The reciprocal of the major rectangle is accomplished by locating the point where the diagonal of the major rectangle and the inside edge of the true square intersect [Hambidge 1960, 28–29; 1967, 35, 60–72].

How to apply areas to a root-two rectangle

- Locate the root-two rectangle (DBEF) of sides 1 and √2, and its diagonal BF.
- From point B, draw a quarter-arc of radius BD that intersects line BE at point K.
- From point K, draw a line perpendicular to line BE that intersects line FD at point L.

The result is a square (DBKL) "applied" on the short side (DB) of the root-two rectangle (DBEF).

- Locate the diagonal BF of the root-two rectangle (DBEF).
- Locate point M where the diagonal (BF) intersects line KL.
- Draw a line through point M that is perpendicular to line KL and intersects line DB at point N and line EF at point P.
- Locate the rectangle NBKM.

The rectangle (NBKM) is the reciprocal of the root-two rectangle DBEF. Line BM is the diagonal of the reciprocal (NBKM) (fig. 13.)

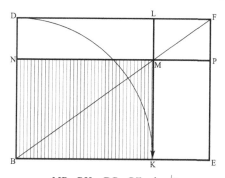

NB : BK :: DB : BE :: 1 : √2
BM : BF :: 1 : √2

Fig. 13   Fig. 14

- Locate the square (DBKL) and the rectangle BEPN.

The area of the rectangle (BEPN) and the area of the square (DBKL) are equal. In Hambidge's system, the "square" is applied on the short and long sides of the root-two rectangle (fig. 14).[10]

- Locate the reciprocal NBKM of the major root-two rectangle DBEF.
- Locate the rectangles DNML and KEPM.

Rectangle DNML is the complement of the reciprocal NBKM.[11]

The areas of rectangles DNML and KEPM are equal.

- Locate rectangles NBKM and LMPF.

The rectangles NBKM and LMPF share the same diagonal and are similar (fig. 15).[12]

Any quadrilateral figure can be applied to the long and short sides of a root rectangle in this fashion. In figure 16 the areas of rectangles DBQR and BEUT are equal and the areas of rectangles DTSR and QEUS are equal.

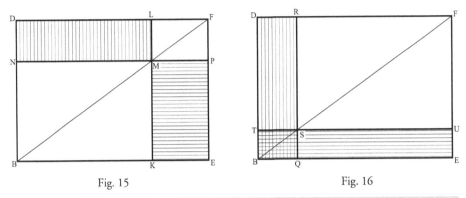

Fig. 15   Fig. 16

Definitions:

Hambidge identifies three ways to apply one area to another. The applied area can be less than, equal to, or in excess of the other. If the applied area is less, it is **elliptic**; if equal, it is **parabolic**; and if in excess, it is **hyperbolic**.

If a square is applied to the short side of a rectangle, it "falls short" and is elliptic. If a square is applied to the short side of a root-two rectangle, the excess area contains a square and a root-two rectangle (fig. 17a). If squares are applied to both short sides of a root-two rectangle, they overlap and the total rectangle divides into three squares and three root-two rectangles (fig. 17b). If a square is applied to the long side of a rectangle, it "exceeds" the rectangle and is hyperbolic. If a square is applied to the long side of a root-two rectangle, the excess area contains two squares and a root-two rectangle (fig. 17c).[13]

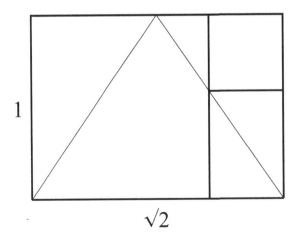

1

$\sqrt{2}$

Fig. 17a

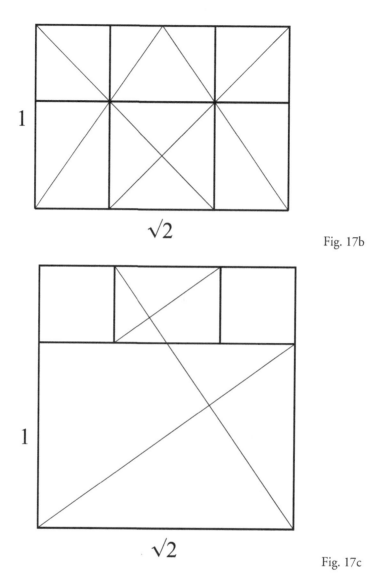

Fig. 17b

Fig. 17c

The Root-Two Rectangle and Variations on Area Themes

The application of areas permits the division of root rectangles into harmonious compositions that are based on the ratio of the overall figure. Let us consider the root-two rectangle and related figures.[14]

How to divide a square into root-two proportional areas

- Draw a square (DBAC) of side 1. (Repeat figures 2 and 3.)

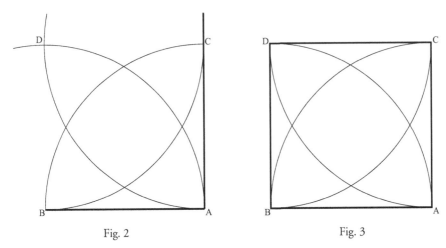

Fig. 2   Fig. 3

- Draw the diagonals AD and CB through the square (DBAC).
- Locate point O, where the two diagonals intersect.
- Place the compass point at A. Draw an arc of radius AO that intersects line AC at point E.
- Place the compass point at B. Draw an arc of radius BO that intersects line BD at point F.
- Connect points E and F.

The result is a root-two rectangle (AEFB) of sides $1/\sqrt{2}$ and 1 (0.7071... and 1) (fig. 18.)

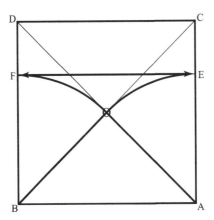

Fig. 18

- Place the compass point at A. Draw a quarter-arc of radius AO that intersects line AC at point E and line AB at point G.

- Place the compass point at B. Draw a quarter-arc of radius BO that intersects line BD at point F and line BA at point H.
- Place the compass point at D. Draw a quarter-arc of radius DO that intersects line DC at point I and line DB at point J.
- Place the compass point at C. Draw a quarter-arc of radius CO that intersects line CD at point K and line CA at point L.
- Connect points J and L. Connect points I and H. Connect points G and K.

The results are three root-two rectangles (CKGA, DJLC, and BHID). When root-two rectangles are drawn on all four sides of a square, the result is a composition that contains one center square, four smaller corner squares, and four root-two rectangles (fig. 19).[15]

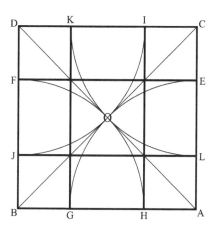

Fig. 19

How to divide a root-two rectangle into segments that progress in the ratio 1 : √2

- Draw a root-two rectangle (DBEF) of sides 1 and √2. (Repeat figures 2, 3, 7, and 8.)

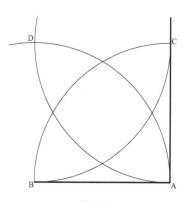

Fig. 2

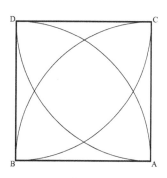

Fig. 3

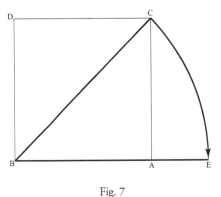

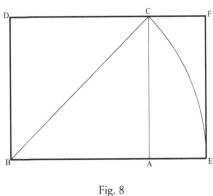

Fig. 7   Fig. 8

- Draw the diagonal (BF) of the major rectangle (DBEF), the reciprocal HEFG, and its diagonal (EG). (Repeat figures 9 and 10.)

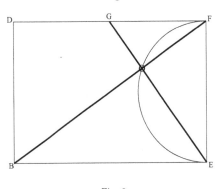

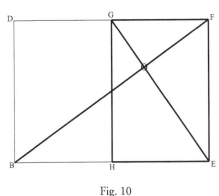

Fig. 9   Fig. 10

The diagonal (BF) of the major rectangle (DBEF) and the long side (GH) of the reciprocal (HEFG) intersect at point I.

- Connect points G and I.
- From point I, draw a line perpendicular to line IG that intersects the diagonal (EG) of the reciprocal (GHEF) at point J.
- From point J, draw a line perpendicular to line JI that intersects the diagonal BF at point K.
- From point K, draw a line perpendicular to line KJ that intersects the diagonal EG at point L.

The process can continue infinitely.

The lines LK, KJ, JI, IG, GF, FE, and EB compose a rectilinear spiral that increases in root-two proportion and decreases in the ratio of 1 : 1/√2 toward the pole or eye (point O) (fig. 20).

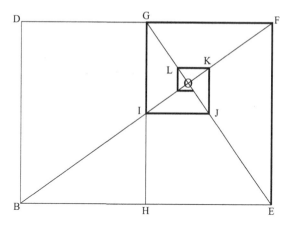

KJ : JI :: JI : IG :: IG : GF :: GF : FE :: 1 : √2

Fig. 20

- Locate the root-two rectangle (DBEF) of sides 1 and √2.

- Draw the diagonals (BF and DE) of the root-two rectangle DBEF, the diagonals (EG and HF) of the reciprocal HEFG, and the diagonals (HD and BG) of the reciprocal BHGD (fig. 21).

* * *

Use the diagonals to repeat the equiangular spiral three times, as shown (fig. 22).

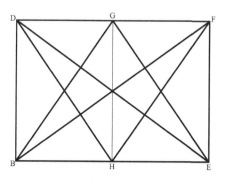

Fig. 21

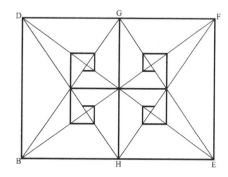

Fig. 22

How to divide a root-two rectangle into smaller root-two rectangles of equal area

- Locate the root-two rectangle (DBEF) of sides 1 and √2.
- Locate the diagonals (BF and DE) of the root-two rectangle DBEF.

- Locate the diagonals (HD and BG) of the reciprocal root-two rectangle BHGD and the diagonals (EG and HF) of the reciprocal root-two rectangle HEFG.
- Locate the midpoints (G, H, M, and N) of the root-two rectangle DBEF.
- Connect midpoints G and H, then midpoints M and N.

The result is a root-two rectangle divided into four smaller root-two rectangles of equal area (fig. 23).

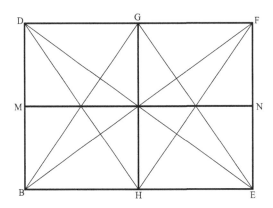

Fig. 23

- Locate points O, P, Q, and R, where the diagonals intersect, as shown.

- Through these points, extend two vertical and two horizontal lines to the sides of the original rectangle, as shown.

The result is a root-two rectangle divided into nine smaller root-two rectangles of equal area (fig. 24).

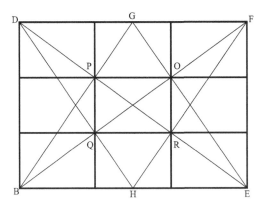

Fig. 24

- From midpoint M, draw the half diagonals ME and MF through the root-two rectangle, as shown.
- From midpoint N, draw the half diagonals ND and NB through the root-two rectangle, as shown (fig. 25).

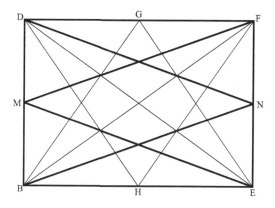

Fig. 25

- Locate points S, T, U, and V, where the diagonals and half diagonals intersect, as shown.
- Through these points, extend three vertical and three horizontal lines to the sides of the original rectangle, as shown.

The result is a root-two rectangle divided into sixteen smaller root-two rectangles of equal area (fig. 26).

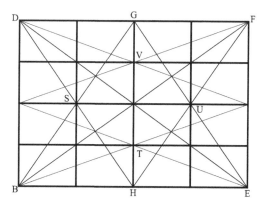

Fig. 26

- Extend four vertical and four horizontal lines to the sides of the original rectangle, through the points of intersection, as shown.

The result is a root-two rectangle divided into twenty-five smaller root-two rectangles of equal area (fig. 27).[16]

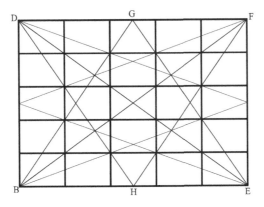

Fig. 27

How to divide a root-two rectangle into patterns of proportional areas

- Locate the root-two rectangle (DBEF) of sides 1 and √2.
- Apply a square (DBAC) to the left side (DB) of the root-two rectangle, as shown (fig. 28).

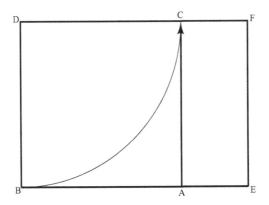

DB : BE :: 1 : √2

Fig. 28

- From the midpoint G of line FD, draw the half diagonals GB and GE through the root-two rectangle, as shown.
- Locate point W where the half diagonal GE intersects the right side (AC) of the square (DBAC), as shown.
- From point W, draw a line perpendicular to line AC that intersects line EF at point X.

When a square (DBAC) is "applied" on the short side of a root-two rectangle (DBEF), it falls short and is elliptic. The excess area (AEFC) is composed of a square (WXFC) and a root-two rectangle (AEXW) (fig. 29).

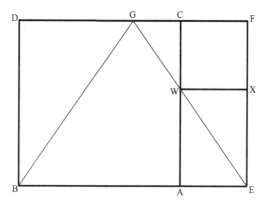

Fig. 29

- Apply a square (EFYZ) to the right side (EF) of the root-two rectangle, as shown (fig. 30).

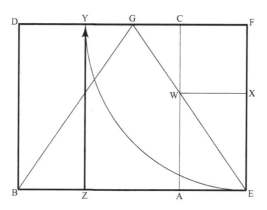

Fig. 30

- From point D, draw the diagonal DA of the square DBAC. From point F, draw the diagonal FZ of the square EFYZ, as shown.
- Locate point A' where the diagonal DA and the half diagonal GB intersect, as shown.
- Extend the line XW through point A', as shown, until it intersects the line DB at point B'.

When squares (DBAC and EFYZ) are "applied" on both short sides of a root-two rectangle (DBEF), they overlap. The various diagonals reveal a pattern of three squares and three root-two rectangles (fig. 31).

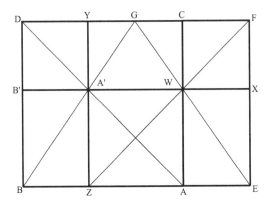

Fig. 31

- Locate the root-two rectangle (DBEF) of sides 1 and √2.
- Locate the squares DBAC and EFYZ applied to the root-two rectangle (DBEF), as shown.
- From point B, draw the diagonal BC of the square DBAC. From point E, draw the diagonal EY of the square EFYZ, as shown.
- Locate point C' where the two diagonals intersect (fig. 32).

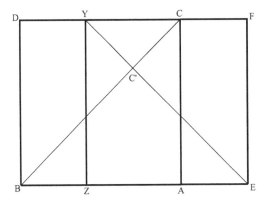

Fig. 32

- Draw a line through point C' that is perpendicular to and intersects line FD at point G and line BE at point H.
- Draw a line through point C' that is perpendicular to line GH and intersects line DB at point D' and line EF at point E' (fig. 33).

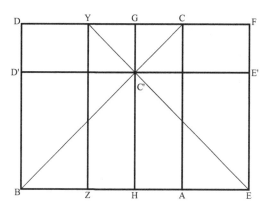

Fig. 33

- From point D, draw the diagonal DA of the square DBAC. From point F, draw the diagonal FZ of the square EFYZ, as shown.
- Locate point F′ where the two diagonals intersect.
- Draw a line through point F′ that is perpendicular to line GH and intersects line DB at point G′ and line EF at point H′.

The four diagonals reveal a pattern of six squares and six root-two rectangles (fig. 34).

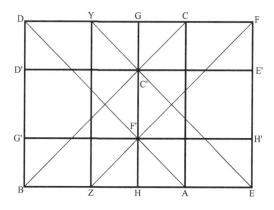

Fig. 34

- Locate the root-two rectangle (DBEF) of sides 1 and $\sqrt{2}$.
- Locate the square DBAC applied to the root-two rectangle (DBEF), as shown.
- From point B, draw the diagonal BF of the root-two rectangle DBEF.
- Locate point I′ where the diagonal BF intersects the right side (AC) of the square (DBAC), as shown.
- Draw a line through point I′ that is perpendicular to line AC and intersects line DB at point D′ and line EF at point E′.
- Locate the rectangle BEE′D′.

The area of the square DBAC and the area of the rectangle BEE′D′ are equal. The "square" is applied on the short and long sides of the root-two rectangle.

- Locate the rectangle D′BAI′.

Rectangle D′BAI′ is the reciprocal of the major root-two rectangle DBEF.

- Locate the rectangles DD′I′C and AEE′I′.

Rectangle DD′I′C is the complement of the reciprocal D′BAI′.

The areas of rectangles DD′I′C and AEE′I′ are equal.

- Locate the rectangles D′BAI′ and CI′E′F.

The rectangles D′BAI′ and CI′E′F share the same diagonal and are similar (fig. 35).

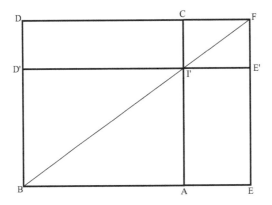

Fig. 35

- Locate the square EFYZ applied to the root-two rectangle (DBEF), as shown.
- Locate the diagonals (BF and DE) of the root-two rectangle DBEF.
- Locate point J′ where the diagonal DE intersects the left side (YZ) of the square (EFYZ), as shown.
- Locate point K′ where the diagonal BF intersects the left side (YZ) of the square (EFYZ), as shown.
- Locate point L′ where the diagonal DE intersects the right side (AC) of the square (DBAC), as shown.
- Draw a line through points K′ and L′ that intersects line DB at point G′ and line EF at point H′.

The diagonals BF and DE of the major root-two rectangle (DBEF) reveal a pattern of six squares and five root-two rectangles (fig. 36).

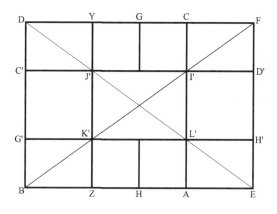

Fig. 36

Design applications

- Locate the root-two rectangle (DBEF) of sides 1 and √2.
- Construct a diagonal grid composed of
  ○ the diagonals of the major root-two rectangle (DBEF)
  ○ the diagonals of the two reciprocals (BHGD and HEFG)
  ○ the diagonals of the two squares (DBAC and EFYZ). (See figure 37.)

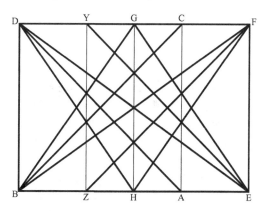

Fig. 37

Figures 38–42 illustrate patterns of proportional areas based on the ratio 1 : √2.

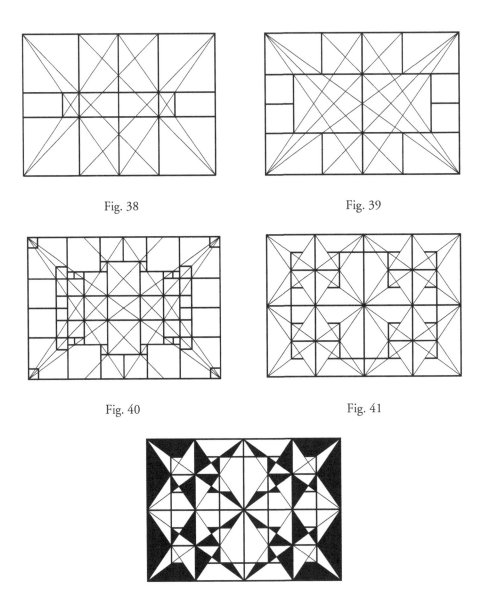

Fig. 38

Fig. 39

Fig. 40

Fig. 41

Fig. 42

Rectangles based on root-two proportions

Besides the square and the root-two rectangle, the designer may use a variety of quadrilateral figures when composing root-two patterns of dynamic symmetry.

The rectangle formed by a square plus a root-two rectangle is in the ratio 1:1 + √2 or 1:2.4142... (fig. 43).

The rectangle formed by the reciprocal of a root-two rectangle plus a square is in the ratio 1 : 1 + 1/√2 or 1: 1.7071... (fig. 44).

The rectangle formed by a square plus the reciprocal of a root-two rectangle on either side is in the ratio 1 : 1 + 2/√2 or 1 : 2.4141... (fig. 45).

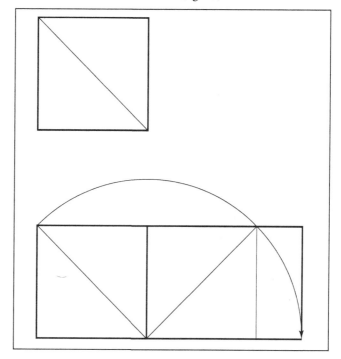

Fig. 43

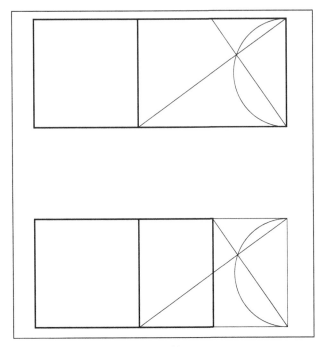

Fig. 44

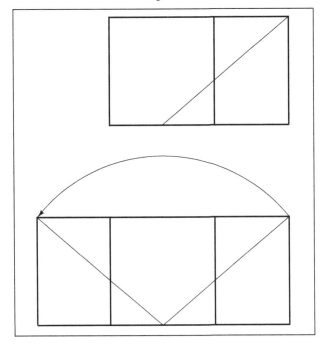

Fig. 45

Conclusion

Ratios inherent in geometric forms offer a rich vocabulary for achieving unity among a diversity of elements. But such systems cannot guarantee good design any more than rules of scale ensure good musical composition. They are mere theories that must be judged by how they serve. No ordering system has merit that does not inform the situation at hand. The capacity of proportional techniques to support and enhance specific cultural and functional requirements remains a matter of individual talent.

Notes

1. "Dynamic symmetry" appears in root rectangles based on square root proportions. The edge lengths of such rectangles are incommensurable and cannot divide into one another. But a square constructed on the long side of the rectangle can be expressed in whole numbers, relative to a square constructed on the shorter side [Hambidge 1960, 22–24; 1967, 17–18]. See [Fletcher 2007] for more on dynamic symmetry and its fundamental components.

2. Hambidge believes the artistic application of dynamic symmetry disappeared after the Classical Greek period of the sixth and seventh centuries BC, but its theoretical expression remained in Euclidean geometry. He speculates that the dynamic proportioning of rectangular forms evolved from the ancient Egyptian practice of "cording the temple" to lay out building plans, including the cord's division into twelve (3 + 4 + 5) equal units to form a right angle [Hambidge 1920, 7–12]. Edward B. Edwards argues that the artistic application of dynamic symmetry persisted well beyond classical Greece, in numerous styles of ornament [1967, viii–ix]. See also [Scholfield 1958, 116–119].

3. See [Fletcher 2007, 329–334] to draw the construction.

4. The construction is taken from [Edwards 1967, 10–11].

5. See [Fletcher 2007, 345–346] to draw the construction.

6. The diagonal is the straight line joining two nonadjacent vertices of a plane figure, or two vertices of a polyhedron that are not in the same face. The reciprocal of a major rectangle is a figure similar in shape but smaller in size, such that the short side of the major rectangle equals the long side of the reciprocal [Hambidge 1967, 30, 131]. See [Fletcher 2007, 336] and [Hambidge 1967, 33–37] for more on these elements.

7. A root-two rectangle is in the ratio 1 : 1.4142... . Its reciprocal ($\sqrt{2}/2$: 1) is in the ratio 0.7071... : 1. A root-three rectangle is in the ratio 1 : 1.732... . Its reciprocal ($\sqrt{3}/3$: 1) is in the ratio 0.5773... : 1. A root-four rectangle is in the ratio 1 : 2.0. Its reciprocal ($\sqrt{4}/4$: 1) is in the ratio 0.5 : 1. A root-five rectangle is in the ratio 1 : 2.236... . Its reciprocal ($\sqrt{5}/5$: 1) is in the ratio 0.4472... : 1.

8. For more on the pole or eye, see [Fletcher 2004, 105–106].

9. The construction applies Euclid's method of obtaining the mean proportional of two given lines [1956, II: 216 (bk. VI, prop. 13)].

10. For the geometrical proof, see [Hambidge 1967, 46].

11. In Hambidge's system, each rectangle has a reciprocal and each rectangle and reciprocal have complementary areas. The complementary area is the area that remains when a rectangle is produced within a unit square. If the rectangle exhibits properties of dynamic symmetry, its complement will also. See [Fletcher 2007, 347–354] and [Hambidge 1967, 128].

12. Rectangles are similar if their corresponding angles are equal and their corresponding sides are in proportion. Similar rectangles share common diagonals.

13. See [Hambidge 1920, 19–20; 1960, 35].

14. See [Hambidge 1967, 40–47].

15. See [Fletcher 2005b, 56–62] for similarities to the Sacred Cut.

16. See [Schneider 2006, 30-32, 34] for similar constructions.

References

EDWARDS, Edward B. 1967. *Pattern and Design with Dynamic Symmetry: How to Create Art Deco Geometrical Designs*. 1932. New York: Dover.

EUCLID. 1956. *The Thirteen Books of Euclid's Elements*. Thomas L. Heath, ed. and trans. Vols. I–III. New York: Dover.

FLETCHER, Rachel. 2004. Musings on the Vesica Piscis. *Nexus Network Journal* **6**, 2 (Autumn 2004): 95–110.

———. 2005a. Six + One. *Nexus Network Journal* **7**, 1 (Spring 2005): 141–160.

———. 2005b. The Square. *Nexus Network Journal* **7**, 2 (Autumn 2005): 35–70.

———. 2007. Dynamic Root Rectangles. Part One: The Fundamentals. *Nexus Network Journal* **9**, 2 (Spring 2007): 327–361.

HAMBIDGE, Jay. 1920. *Dynamic Symmetry: The Greek Vase*. New Haven: Yale University Press.

———. 1919–1920. *The Diagonal*. **1**, 1–12 (November 1919–October 1920). New Haven: Yale University Press: 1–254.

———. 1960. *Practical Applications of Dynamic Symmetry*. 1932. New York: Devin-Adair.

———. 1967. *The Elements of Dynamic Symmetry*. 1926. New York: Dover.

SCHNEIDER, Michael S. 2006. *Dynamic Rectangles: Explore Harmony in Mathematics and Art*. Constructing the Universe Activity Books. Vol. IV. Self-published manuscript.

SCHOLFIELD, P. H. 1958. *The Theory of Proportion in Architecture*. Cambridge: Cambridge University Press.

About the Geometer

Rachel Fletcher is a theatre designer and geometer living in Massachusetts, with degrees from Hofstra University, SUNY Albany and Humboldt State University. She is the creator/curator of two museum exhibits on geometry, "Infinite Measure" and "Design By Nature". She is the co-curator of the exhibit "Harmony by Design: The Golden Mean" and author of its exhibition catalog. In conjunction with these exhibits, which have traveled to Chicago, Washington, and New York, she teaches geometry and proportion to design practitioners. She is an adjunct professor at the New York School of Interior Design. Her essays have appeared in numerous books and journals, including *Design Spirit, Parabola,* and *The Power of Place*. She is the founding director of Housatonic River Walk in Great Barrington, Massachusetts, and is currently directing the creation of an African American Heritage Trail in the Upper Housatonic Valley of Connecticut and Massachusetts.

Jane Burry
Andrew Maher

Design and Social Context
RMIT University
GPO Box 2476V
Melbourne 3001
AUSTRALIA
jane.burry@rmit.edu.au
andrew.maher@rmit.edu.au

Keywords: structural
systems, algorithms,
engineering, geometry,
Euclidean geometry, non-
Euclidean geometry

Didactics

The Other Mathematical Bridge

Abstract. This paper contextualises, describes and discusses a student project which takes a particular exploratory approach to using mathematical surface definition as a language and vehicle for co-rational design co-authorship for architecture and engineering. The project has two authors, one from an architectural and one from an engineering educational background. It investigates the metaphorical and operational role of mathematics in the design process and outcomes.

1.0 Introduction

1.1 Architecture and engineering

Within design, there appears to be, once again, a keen interest in a type of organicism that emerges from an underlying 'system', complexity from simple roots, form that follows certain growth criteria and responds 'intrinsically' to constraint systems such as gravity and site conditions or the external forces of weather and use. Perhaps this is to be expected in the time of the fine-grain decoding of life as DNA.

This appears fertile philosophical soil in which the curiously estranged disciplines of architecture and structural civil engineering can grow together. Accepting that the coexistence of these two fields with apparently similar objectives has a historical foundation, it is clear that the points of distinction have not remained consistent throughout history. Robin Evans notes a passing of the baton in late eighteenth century as descriptive geometry, specifically stereotomy, a minority but virtuoso technique in architecture, was passed, in the influential writing of Monge, from the architect and stonemason to the engineering community, to be taken up in their new roles as designers of steam ships and locomotives in the nineteenth century [Evans 1995: 328]. The means of representing and describing spatial conceptions, the particular preoccupations with materials, the types of value ascribed to various attributes of the design within architecture and engineering may always have followed separate paths but these wandering paths have crossed.

1.2 Architecture without mathematics

To some extent the particular disciplinary approaches to description can be said to have fed back into the conceptual process of form making. It is not hard to defend the observation that many architectural education courses have progressively eschewed any interest in the deployment of formal mathematics. Spatiality and spatial organisation may be innately mathematical but consideration of proportion, statics, and the manual construction of perspective views have now largely left the pedagogy, following mathematical description of surface, now long departed. Projection remained (and largely remains) an important conventionally prescribed conceit that bears heavily on the conceptualisation and realisation of architecture, but the strict Cartesian stranglehold on design has been relaxed as it has been internalised and obscured in hidden algorithms [Pérez-Gómez and Pelletier 1997: 378]. Now that much of the work of projection has been subsumed by the machine, the allusion to full shifting-perspective occupiable space, freed from its three imposed axes and fixed view points, makes formal and spatial complexity outside a single framework more accessible. The orthogonally gridded world may still be the most prevalent procurement reference frame but conceptual spatial design need not set up its relationships according to this universal locator. The whole of Euclid is now not only available as a conceptual framework (as it always was) but is relatively effortless to deploy; manifolds that exhibit non-Euclidean characteristics at large scale are also within reach of the three dimensional virtual modeller. 'Digital clay' is still relatively geometrical or, at its most analogous, still influenced by the particular surface algorithms available. However, it is possible, with a little effort, to work in earthen clay or plaster[1] and find a geometrical description or controlled surface rationalisation method later through semi-automated processes that help realise tactile scale models at architectural scale and complexity.

Contemporary conceptual design space in education is potentially a no rules space or, at least, the search for appropriate rules systems has opened up with the technological means to appropriate more complex geometrical structures and programs from other fields. In place of truth, designers seek and use what is productive, and enjoy dialectics. Aesthetic arbitration holds less interest than defining the framework from within which it is being exercised. Liberated from any one universal constraining context, designers can choose their tools and their goals for their exploration of spatial possibilities. Some exercise this choice.

1.3 Structural economy

While the means to explore structural economy are ever more computationally sophisticated, this cornerstone of modernist ideology may be seen to have slipped from architectural prominence. The idea of a graph of relative use of steel by weight against the covered area of new Olympic stadia since Günther Behnisch and Frei Otto's revolutionary lightweight proposal in 1972 would be instructive in this regard.[2] In main stream practice, economy may commonly be met by reducing the overall number of standardised structural members used or finding an effective mean span but there are more subtle approaches to conceptual structural design that may also align with biological metaphor. D'Arcy Wentworth Thompson provides explanations for the scale and form of the living, demonstrating the diversity and specificity of evolutionary outcomes all conforming to the same Newtonian physics [Thompson 1992]. Cell growth too is a stimulating, potentially useful metaphoric process to consider in relation to *designing the design* for structural systems, conforming as it does to the genetic blueprint while, at a micro level, cells are laid down and removed in response to local structural exigency. What is the computer for if not to test these ideas by simulating, or at least emulating, the binary aspects of such processes to find structures that obey the same basic principals of getting the best performance for the least? What are the aesthetics of this kind of minimalism?

1.4 Form finding systems

The terms 'form finding' and 'generative design' refer in a general way to investing creatively in the design of a system to define possible formal/spatial outcomes according to specific relationships and criteria rather than in the more deterministic design of specific spaces or forms.[3] They are to some extent generic descriptions of several specific processes with different objectives. One manifestation is a graph of geometrical relations that supports a consistent topology that can generate many different contextually responsive forms and selecting forms from this field of possibilities through optimisation for certain ranges of values of particular parameters or relationships between them. An example of this might be a roofing element with sculptural qualities that admits indirect light into the building applied in a grid across a changing undulating roof structural system and optimising the roof form so that the largest percentage of the rooflights are within a value of a particular compass point. This might yield a range of solutions better and worse for different reasons. The actual geometry and dimensions of every roof element can be unique while conforming to the topological blueprint and recognisably similar across the field of instantiations, much as individuals vary in a population of oysters across an oyster bed. This looks like a return to highly Platonic concept of a contingent perceptual world of (imperfect, imprecise) copies of ideal forms.

Another specific and contrasting example is the deterministic optimisation method called 'Evolutionary structural Optimisation' [Xie and Steven 1997: 97]. This is an iterative structural optimisation tool, closely analogous to processes in nature. It uses finite element analysis to identify and remove the least stressed material in a structure. This process is repeated many times and a highly structurally optimised form emerges. The form is determined by the way that the loads have been applied and the way the object is supported. Only one particular optimised form can be found within a particular set of conditions. A simple and classic example shown to illustrate the process by Prof Y. M. Xie is the evolution of a cube suspended from one central top point. At the end of the process it appears the shape of an apple. In reality this method has been developed to generate complex three-dimensional structural forms, the current version allowing both the subtractive and additive processes in response to both compressive and tensile stresses. This process is essentially non-geometrical by its treatment of structure as finite elements.

2.0 The project

2.1 Context of the project

Form finding lies within the broad territory negotiated between architecture and structural engineering where the paths are likely to cross. Potentially this crossroads should be a most fruitful and emergent social and operational locale for conceptual design activity for built systems. *Dissolving the Boundaries between Architecture and Engineering* is the name of one of the eleven research projects under the broad umbrella of the Virtual Research and Innovation Institute (VRII) for Information and Communication Technology (ICT) at RMIT University to research ways to broaden this shared activity. It brings together the Spatial Information Architecture Laboratory from the School of Architecture and Design and the Innovative Structures Group from the School of Civil and Chemical engineering, from different faculties.

Although the two professions work together continually, in practice there is a deep cultural and epistemological chasm running between the disciplines that is established and maintained in the education system. This is not limited to any particular university or even country – it appears common, for instance, to Australia, New Zealand, USA, Canada and UK. The institutions where this is patently false are the exceptions in these fields. The gulf may be linked to or exacerbated by the nature of accreditation by the respective professional bodies. As part of the research into dissolving these boundaries a research–based, experimental joint design studio for final year undergraduates from both disciplines was proposed and given the title Re-engineering. This was a loosely structured research project supported by a team of staff from both disciplines with the overriding and much emphasised brief to explore the concept of *co-rational design*. This is the idea that there is a third way as an alternative to either taking a structural system as a point of departure – the *pre-rational* approach – or designing a structure in response to a pre determined formal solution – the *post-rational* approach – or '*please make it stand up*'. The studio provided an environment in which to explore structural systems in synthesis with other design drivers. Each architecture student was paired with an engineering student. One partner (the architecture student) had four years experience of being immersed in a progressively more student-led vertical studio context continually challenged to initiate conceptual design and speculate, all projects focused around the built environment. The other (the engineering student) had been trained for a similar length of time to be a focused problem solver, seeking appropriate solutions to problems posed in a range of engineering contexts of which building structures was one, reporting rigorously on the outcomes of applying solutions and accustomed to a well-defined problem as a starting point.

2.2 Educational methodology

The architecture students were enrolled in a course that requires a speculative semester of supported research in preparation for their Major (or thesis) design project. The engineering students were enrolled in an investigation project, their final project leading to the submission of a written research report. There was no proscription as to the means of communication and sharing, but there was a heavy emphasis placed on the co rational design objective of finding a means to co–authorship. The studio was uncompromisingly process driven and divided into three phases. The three stages were articulated to encourage continual return to the origin throughout the semester, albeit with a more developed focus each time, and to suppress the inclination to develop designs more fully or diverge into separate disciplines in the process. These courses were chosen as the setting for the experiment in collaboration for their emphasis on research. They are open ended, requiring the engineering students to prepare a formal written research report, the architecture students, a more graphical representation of their research findings and their design intentions. In distinguishing between method and methodology, this is action research, generating knowledge through design, by working together on projects. The central research question for all the students concerned a third way of designing that was neither formally reactive to a structural approach nor structurally reactive to a formal approach. The sub-questions were to investigate methods of 'labour' that supported early "*collaboration*" between the designers of the two disciplines and how this collaboration impacted on the quality of their design outcomes.

The weekly classes were structured as a space to present and reflect collectively on the week's work, exploiting 'the mind's ability to ponder its own reflections' (Locke, quoted in [von Glasersfeld 1991: xviii] and to develop 'consensual domains' (Piaget, quoted in [von Glasersfeld 1991: xvi] in a mediated conversation between the students engaged actively in design and staff as research facilitators, critics and commentators.

2.3 The project

Here we chart the development and outcomes of one particular architectural research project undertaken in the context of this course, curious to locate this with regard to the multifarious relationship between mathematics and architecture. In the project described, the overall objective of finding a co rational way to work together conceptually was most successful in the intense closing stages of the project.

The first phase was a series of weekly investigations, each starting with a new program and structural concept chosen or devised by the students. The ground work for this project was wide ranging. First, experimental use of the 2D Evolutionary Structural Optimisation (ESO) tool developed by Prof Mike Xie provided structurally optimised cross sections for an extruded or extrapolated 'Hanging Tower' followed by calibration of the tool by reverse engineering the cross section of Frank Lloyd Wright's Falling Water in order to move on and find a series of varying structurally optimised cross sections responding to local support conditions for the free form envelope of a suspended space in a Melbourne laneway. In subsequent weeks they further experimented with the laneway proposal, developing an undulating shell structure attempting to minimise the surface area and to learn how to use finite element analysis to resolve this into a compression structure, in a way analogous to the funicular model used by Gaudí to find the lines of force and hence the form for the church for the Colonia Güell [Bassagoda Nonell 1989]. They modelled it using the vacuum former and applied physical loads to measure the deflection. They also considered material strength testing.

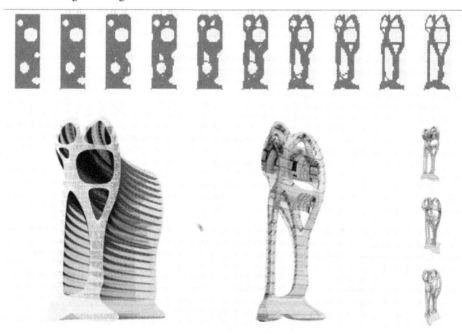

Fig. 1. 'Hanging Tower' developed from a swept cross section optimised for structural performance using a prototypical 2D version of the Evolutionary Structural Optimisation software. Initial exploration of the potential and use of the software by Steven Swain, RMIT architecture pre-major student 2005

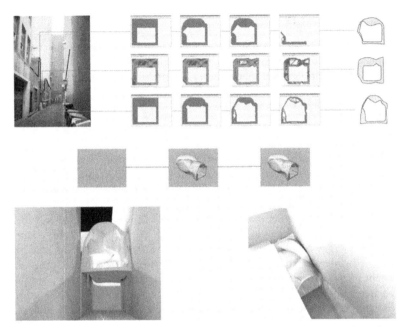

Fig. 2. China Bar showing the same ESO tool used to define a surface from differentiated optimised structural sections for changing support conditions along its length. Joint project by Steven Swain, RMIT architecture pre-major student 2005 and Andew Rovers RMIT final year civil engineering student 2005, in which they also investigated generating catenary forms and framed, tensile and shell structures. They tested a shell structure in structural analysis software and using physical weights on a vacuum-formed model. (Images shown prepared by Steven Swain)

The second phase narrowed the focus to one particular structural approach, possibly one of the first phase experiments. The partnerships changed. The architecture student considering shells and continuous surfaces was now teamed with an engineering student who saw his own strength in mathematical understanding. The 'architect' immediately adopted some difficult, mathematically-derived surfaces. A number of equations were selected including a combined Jacobi elliptic function and hyperbolic cosine function. They were chosen from a library of surfaces on criteria of aesthetics and spatial potential. Through very simple manipulation of these 'found' surfaces – Booleans to create edge boundaries and openings and differential scaling – a series of formal articulations of program were suggested, an interpreted railway station roof, a sinuous tower development. The most compelling was the use of a surface in its most raw state as the shell structure of the *Hybrid Cathedral*. In this proposal the surface mediated between a soaring sacred space of monumental proportions at the heart, and multilevel apartments nestled in the sinuous peripheral undulations. It was prototyped in wax to enjoy its engaging formal-programmatic encounter at a more sensory level. At this point, the differences between the software and processes introduced by the architects and engineers became very apparent and divergent. Apart from the usual issues of format compatibility and transfer, the rigid 'yes or no', 'right or wrong' solution inherent in the engineering software compared to the forgiving nature of the architectural modelling software in supporting speculation, meant the engineer struggled at this stage to define a role.

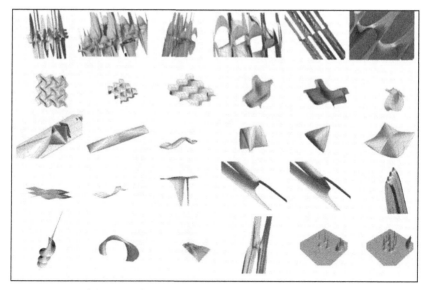

Fig. 3. One of the pages from the collected 'library of surfaces'. Steven Swain (final year architecture) and Sean Ryan (final year civil engineering) experimented with shell structures based on surfaces from a gamma function. Sean researched the function, varied the coefficients in Maple and exported geometry files

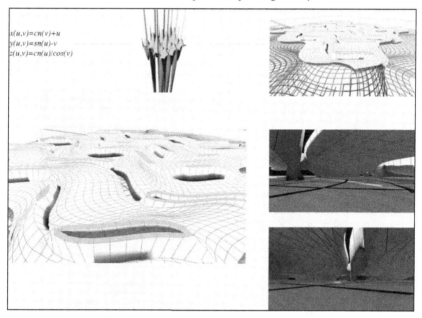

Fig. 4. 'Functional' Railway Station roof using Booleans and scaling to create a programmatic surface from a mathematically defined surface. (Design and images Steven Swain)

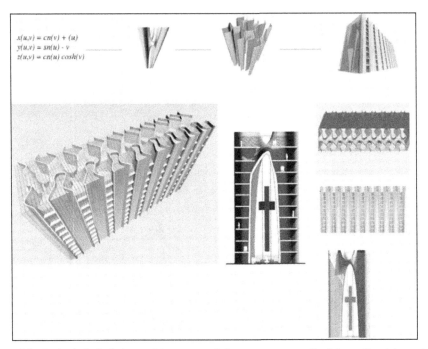

$$x(u,v) = cn(v) + (u)$$
$$y(u,v) = sn(u) - v$$
$$z(u,v) = cn(u) \cosh(v)$$

Fig. 5. The Hybrid Cathedral: worship space and the apartments to fund it mediated by a single mathematical surface. It was proposed for an environment such as Hong Kong, with scarce land and burgeoning population and economy; collaborative project between Steven Swain (architecture) and Sean Ryan (engineering). Steven developed the spatial and programmatic concept, they both worked on refining the surface and Sean endeavoured to undertake finite element analysis of the structural shell and then design a discrete frame structure for analysis in more familiar software

Fig. 6. View of the interior of the Hybrid Cathedral

1.3 The final stage of the project

The third phase introduced site and location, subtly inverting the program-seeking form experimentation earlier. It led this same partnership to the development of an inhabitable bridge that could be mathematically defined. Architectural and structural parameters were identified as they embarked on writing an equation that would satisfy both parties and the program that they had jointly defined. At this stage, a much more intense interaction with the mathematics unfolded.

A publicised but short lived proposal to divert the Geelong freeway across the entrance to Corio Bay was reawakened to advance the concept of a single mathematically controlled surface as structure, rich space defining boundary and interface between monumental scale and domestic infill. The real world requirements of maintaining and spanning the dredged shipping channel, also allowing small craft to pass between Corio Bay and its parent Port Philip Bay, maintaining the tidal flow, observing the spatial, gradient and curvature constraints of the freeway and separating the habitation with its services and access roads from the freeway, intensified the quest to develop the relationship with the surface equation that would allow detailed manipulation of the parameters without relinquishing the emergent qualities and aesthetic coherence of the surface itself. It introduced all the architectural dialectics around the intensity of the experience of crossing the bay at the historic fording point and the iconic and environmental impact of the bridge as it reshaped the view and context for Geelong. It also engaged with the specific engineering challenges of exceptionally long spans, building in water, and site conditions at the springing points.

Simply editing the variables within the original function had a similar impact to scaling the surface using external software algorithms; for instance, it altered the distribution of bridge piers but continued to create repetitive, regularly-spaced piers. To be able to create the large opening for the shipping canal but find more optimised structural intervals for the other parts of the bridge it would be necessary to add a second function to disrupt the rhythm. Various functions were overlaid, some causing too much disruption and surface distortion. Finding a satisfactory addition through empirical experimentation imbued a situated awareness of the power of superposition of different functions, and it was possible in the same way to overlay a fine grain to the surface, a detailed level of surface undulation or corrugation for combined aesthetic and structural opportunity. The formula was then simplified in experimentation to find out how to control the level of detail and hierarchy of peaks, calibrating it to control the height of the peaks in the undulating surface (varying this in relation to the width of the bridge and spans) and a further function superposed to vary the height of these peaks. By this stage, the designers had entered or immersed themselves in equation or function building as their design environment. At each iterative step, the formal elegance and subtlety of the model increased with the increasing control and mastery over its potential to vary. In order to curve the bridge in plan into the sinuous 'S' needed to meet the freeway routing at each abutment, and make the crossing at the old fording route, some of the existing components of the function could be used but had first to be rearranged and separated or their impact altered through denomination. The peaks then had to be controlled in a way that specifically reduced their height at the springing, where short piers were required, and at the main shipping canal, where the vast span would require stiffness but all possible reduction in the weight. This variation could be periodic but the period relative to the pier intervals needed to be controllable in a specific way. This required the further superposition of a specific function for u and v in Z. Although the formal mastery now extended to understanding how to vary not simply the parameters but the function itself, the means to arch the bridge deck following a specific curve from springing to springing was not yet clear.

The source of the original kernel of the function and surface led to the Astro Physics department of Swinburne University. The function was rewritten in a way that clearly parametricised it for the variables already identified and an additional Gaussian function now gave the arch to the road to allow it also to rise up 70m over the shipping channel from its low lying springing points.

The designers could now rewrite the equations satisfied by the x, y and z values of each u, v point on the surface with the list of variables in Table 1.

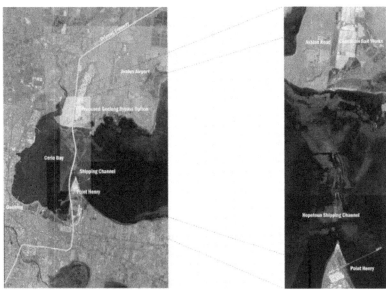

Fig. 7. The Geelong Bypass bridge site. The bridge proposal in this project was to be a 5-kilometre long, one-street highway town, with prime views and real estate helping to fund its construction. The freeway and bridge city were to be separated by a surface generated to test a particular mathematical function. Ferry terminals at the base of the piers allow the residents to reach Melbourne and Geelong without entering the freeway. This was the final collaborative project by Steven Swain (architecture) and Sean Ryan (engineering); image by Steven Swain

Variable parameters in bridge surface function
1   Number of piers
2   Height of the piers
3   Width of the road
4   Length of the road
5   Number of cycles in the x-y plane (controlling the plan curvature of the road)
6   Amplitude of cycles in the x-y plane (also controlling the plan curvature of the road)
7   Number of cycles in the secondary function controlling the varying heights of the piers
8   Amplitude of the this secondary function
9   Height of the road arch (number of cycles will be constant for the single arch)

Table.1

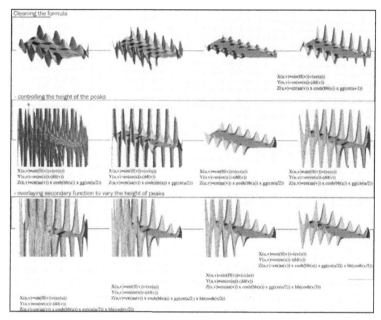

Fig. 8. Manipulation of the function showing the effects of superposition of functions, restructuring the function and parameter value changes. (Project: Steven Swain and Sean Ryan, images: Steven Swain)

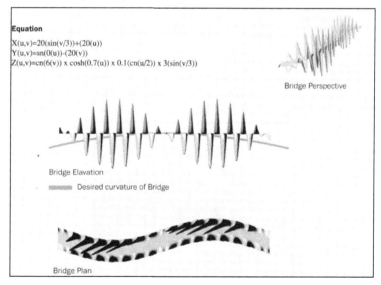

Fig. 9. The bridge surface showing its potential for manipulation for all required characteristics except achieving the necessary arch from its springing up seventy metres over the shipping canal without causing surface distortion. (Project: Steven Swain and Sean Ryan, images: Steven Swain)

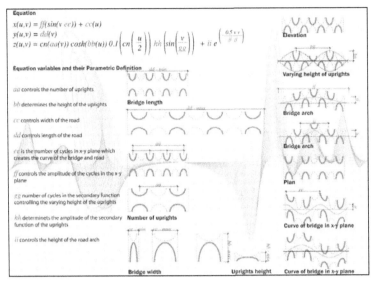

Fig. 10. Illustrated table of variable parameters for the definition of the bridge form. At this stage the architecture and engineering collaborators reported having a medium in which there were able to negotiate the design inputs, overcoming some of the earlier frustrations of the engineering student who felt at the second stage, that, unless detailed structural analysis was called for, he had little input. (Project: Steven Swain and Sean Ryan, images: Steven Swain)

Fig. 11. One of the alternative experimental iterations of the bridge. (Project: Steven Swain and Sean Ryan, images: Steven Swain)

Fig. 12. View of the bridge in its final form. (Project: Steven Swain and Sean Ryan, images: Steven Swain)

A lower deck was needed below the freeway to provide access to the inhabited pier shells. For this the same functions could be used with a small change to the value of the last variable: altering one variable in the short Gaussian expression.

In summary every aspect of the bridge is periodic, determined by its tidy three line function but the superposition of these periodic 'behaviours' is formally subtle and variably aligned with programmatic constraints. Its description is simple and simply conveyed or *transmissible*; its spatial manifestation rich and animalistic.

3.0 Process, outcomes and reflections

While the ultimate requirement for a formal written report as the deliverable from the engineering students left much of the best graphical illustration of the projects in the portfolios of the architecture students, it provided more insightful written reflections than were garnered from the students' comments in weekly class discussions. The engineering authors were honest in expressing frustration when paired with an architecture student they regarded as less than conscientious or productive but also inferred that it was part of the learning process to become accustomed to other working styles. In their literature reviews they gathered strong evidence for the value of a 'co-rational' approach to architect-engineer design and were initially optimistic that this could be realised through a will to work together. They reported very successful outcomes at the end of the first project. One of the engineering students listed the criteria of success that he intuited by the end of the first project as "Understanding of Language, Understanding of Capabilities, Uncertainty, Data (interoperability) and Attitude (levels of openness and interest)". The same author was amazed to discover the differences in the common use of language by the two disciplines, the time taken to overcome this and the dangers inherent in trying to adopt terms one from the other. He observed that on each side the capability expectations were extremely high – he had assumed that architecture trained students would quickly and naturally create beautiful forms that worked and have these modelled rendered in 3D in no time while his architecture counterparts assumed there was nothing much to analysing a structure and expected instant results. They experienced software interoperability challenges and more fundamental ontological differences between softwares conceived to be speculative and those for analysis with little room for fallibility.

By the end of the second project, they reported that, in general the partnerships had failed to achieve a 'co-rational' result. They attributed this to different causes: lack of time and opportunity to work together in the same space, interoperability challenges, poor level of interest from their partners in the particular challenges of their own discipline, working with new techniques beyond current capabilities, but the result seemed to be the same – the retreat of each partner to familiar territory within their own disciplines.

At the end of the third project, the view was expressed that "it is very difficult to develop or even define a formal co-rational design approach". They attributed this to a fundamental distance between architecture and engineering typified by respective emphasis on art and science and reinforced in the pedagogy. At the same time they reflected on the transformative nature of the experience which had given them a much greater appreciation of the work of the architect, greatly expanded their geometrical repertoire and interest in innovation in structures, enhanced their conceptual design skills and fundamentally changed the way they looked at problem solving and design issues.

It was interesting to read their own perception of *frustration* in the stated objective of designing co-rationally in relation to their preconceptions of how this might be, and contrast this to the collective academic's perception of *success* in relation to the same measure, because the students had found (in the example illustrated, mathematical) media in which to hold a design conversation and develop an outcome that was not clearly spatially or structurally led. In other words, they had uncovered a form of labour that supported *co/laboration*. It would have been interesting to have had equivalent written reflections from the architecture students. One measure of their evaluation of the way in which the collaboration contributed the quality of design outcomes was the verbal concern expressed about having to develop the work for their Major project the following semester without the further input of the engineering students. Another, in the illustrated example, was that work primarily representing research won State and National architectural student awards in design categories.

4.0 Discussion

What is the significance of this work? Clearly in the context of the particular academic studio in which it was taken, its significance lies in this *transmissibility*;[4] its power as a common vehicle for a student of architecture and a student of engineering from their strictly segregated educational cultures to work concurrently on formal conceptual design. 'In the long run what must be transmitted is not the object itself but its cipher, the genetic code for the object at each new site, according to each site's available resources' [Novak 1996].

In the context of architectural borrowing, inheritance, deep inspiration from science and mathematics, what is the significance of the experimental application of the discoveries of Carl Gustav Jacob Jacobi around 1830 and Gaussian number theory developed in the closing years of the eighteenth century in a joint architecture engineering studio in 2005?

Perhaps it is Antoine Picon's hypothesis that it is the similarity of operation between science and architecture that at certain points makes the relationship most productive. Picon and Ponte also write of 'a new type of connection between architecture and science' for which 'the computer, of course, is central' [Picon and Ponte 2002: 14]. The tradition of metaphorical and methodological exchange between science and mathematics and architecture goes back a long way. In the fifteenth century, Alberti took a philosophical and aesthetic lead from the contemporaneous revolution in astronomy and nature's preference for roundness [Wittkower 1952]. The terms 'structures', 'mathematical surfaces', and architectural examination of biological sciences all seem to lead back strongly in our time to the nineteenth century. 'What would nineteenth-century architecture have been without the notion of structure?' [Picon and Ponte 2002: 294]. So analogous to the model for this studio or of design itself, this architecture reaches back in an iterative cycle to retrieve largely nineteenth-century ideas that find new applications underpinned by the current state of technology. Martin Bressani writes that, '[t]he central problem with architecture's relationship with modern science is not the distance that separates the two disciplines but, on the contrary, a closeness that prevents free metaphoric exchanges.' He highlights the nineteenth century as a good illustration of the paradox. The French word 'structure' was first used in biology to denote the internal organisation of the body, and Violet-le-Duc's vast library contained no volume on the modern science of engineering: despite his advocacy of rationalism and structural determinism, his analogies and archaeological methodology were all drawn from physiology, anatomy and geology [Bressani 2002: 120].

What is compelling about mathematical surface definition or generative processes that bear a metaphorical resemblance to the 'laws of nature'? Clearly, there is a rationalist drive to define design objectives as a rule set controlling the configuration of space and form. This is a way to gain greater efficacy from the technology – using computation to achieve a set of complex spatial or geometrical objectives simultaneously through the definition of their relations. Then there is the matter of beauty. There is the rational scientific idea that underlying natural beauty is a profound system of law abiding relationships. By reconstructing a closely analogous system, not only the source but the resulting sensory delight will be rediscovered. Finally there is the distinct question of *mathematical beauty* : the authors' delight in a bridge of great spatial and programmatic complexity from a three line function. This aesthetic is so intensely felt yet so ineffable that even Paul Erdös said on the subject:

> Why are numbers beautiful? It is like asking why is Beethoven's Ninth Symphony beautiful? If you can't see why, someone can't tell you. I know numbers are beautiful. If they aren't beautiful, nothing is [Hoffman 1998: 44].

Acknowledgments

Steven Swain, Architecture RMIT pre-major student in 2005, prepared all the figures presented in this paper from his work for the joint projects with Andrew Rovers and Sean Ryan, final year Civil Engineering students.

RMIT University funded *Dissolving the Boundaries between Architecture and Engineering* as one of the eleven research projects of the Virtual Research and Innovation Institute (VRII) for Information and Communication Technology (ICT) This undergraduate research studio described in this paper was supported by that project.

Dr Saman De Silva and Professor Y. M. Xie, RMIT School of Chemical and Civil Engineering co-taught this course with Jane Burry and Andrew Maher from the Spatial Information Architecture Laboratory, RMIT School of Architecture and Design.

Notes

1. For instance: Frank Gehry's design process. Antoni Gaudí's process for the design and construction of the Sagrada Família church is a good counter-example where ruled surface geometry is used to construct the physical models and also used in their geometrical reverse engineering for continuing design for construction, employing computation for analysis and synthesis.
2. An idea raised by Robert Aish in conversation.
3. For further reading, see [Shea, Aish and Gourtovaia 2003]; [Burry 1998]; [Kolarevic 2003]; [Kilian 2003]; [Xie and Steven 1997].
4. This allusion to the work of Marcus Novak [1996] was included by the student in his project submission.

References

BASSAGODA NONELL, J. 1989. *El Gran Gaudí.* Editorial AUSA: 365-373.

BRESSANI, Martin. 2002. Violet-le-Duc's Optic. *Architecture and the Sciences: Exchanging Metaphors.* Princeton: Princeton Architectural Press.

BURRY, Mark. 1998. Gaudí, Teratology and Kinship. *Architectural Design* issue on 'Hypersurfaces' (April 1998). London: Academy Editions.

EVANS, Robin. 1995. *The Projective Cast: Architecture and its Three Geometries.* Cambridge, MA: MIT Press.

HOFFMAN, Paul. 1998. *The Man Who Loved Only Numbers: the story of Paul Erdös and the search for mathematical truth.* New York: Hyperion

KILIAN, Axel. 2003. Fabrication of Partially double-curved Surfaces out of flat sheet Material through a 3D Puzzle Approach. Pp. 75-83 in *Connecting >> Crossroads of Digital Discourse*, Proceedings of the 2003 Annual Conference of the Association for Computer Aided Design in Architecture.

KOLAREVIC, Branko, ed. 2003. *Architecture in the Digital Age: Design and Manufacturing*. New York & London: Spon Press – Taylor & Francis Group.

NOVAK, Marcus. 1996. Transmitting Architecture: the Transphysical City. In *ctheory.net* (29 November 1996), Arthur and Marilouise Kroker, eds. www.ctheory.net/articles.aspx?id=76.

PEREZ-GOMEZ, Alberto and Louise PELLETIER. 1997. *Architectural Representation and the Perspective Hinge*. Cambridge, MA: MIT Press.

PICON, Antoine and Alessandra PONTE, eds. 2002. *Architecture and the Sciences: Exchanging Metaphors*. Princeton Papers on Architecture. Princeton: Princeton Architectural Press.

SHEA, Kristi, Robert AISH and M. GOURTOVAIA. 2003. Towards Integrated Performance-Based Generative Design Tools. Pp. 553.560 in *Digital Design* (21th eCAADe Conference Proceedings), Graz (Austria) 17-20 September 2003.

THOMPSON, D'A. W. 1992. *On Growth and Form*. New York: Dover Publications.

VON GLASERSFELD, Ernst, ed. 1991. *Radical Contructivism in Mathematics Education*. Kluwer Academic Publishers.

WITTKOWER, Rudolf. 1952. *Architectural Principles in the Age of Humanism*. London: Alec Tiranti.

XIE, Y.M and G.P. STEVEN. 1997. *Evolutionary Structural Optimisation*. Heidelberg: Springer Verlag.

About the authors

Jane Burry is a registered architect and part time Research Fellow in RMIT University's Spatial Information Architecture Laboratory, Melbourne, Australia. In 2007/2008 she holds a visiting position at Queensland Institute of Technology. Her current research focus is the contemporary place of geometrical space in architectural design space. In recent years she has published 25 refereed papers, mainly on the topics of the impacts on the design process of introducing flexible digital parametric modelling in design practice and pedagogy, and on supporting communication in distributed multi-disciplinary design teams.

Andrew Maher is a Research Fellow at RMIT University's Spatial Information Architecture Laboratory, where he investigates technology transfer in architectural practice, including the social and cultural barriers to adoption, and fosters understandings of how new knowledge is created and then 'flows' within these organizations.

Book Review

Warwick Fox

A Theory of General Ethics: Human Relationships, Nature, and the Built Environment

Cambridge, Massachusetts: MIT Press, 2006.

Reviewed by Michael J. Ostwald

School of Architecture and Built Environment
Faculty of Engineering and Built Environment
University of Newcastle
New South Wales, AUSTRALIA 2308
michael.ostwald@newcastle.edu.au

Keywords: ethics, geometry, built environment

When contemporary scholars read and interpret historical architectural treatises, one dimension that is frequently lost from the original works is their moral or ethical values. For example, when John Ruskin argues that there are right and wrong geometric shapes in architecture, he may be talking about the difference between mechanical Euclidean curves and freehand quasi-logarithmic ones, but he is, more importantly, also talking about lines that are not simply geometrically but morally correct as well. Similarly, the texts of Vitruvius and Alberti contain rules for the ideal construction of architectural form, but this rightness is first and foremost a moral or ethical quality that is symbolically embodied in a geometric construction. However, while the geometric construction of historic and modern architecture remains the subject of considerable research today, the ethical dimensions in the same architectural techniques tend to be forgotten or are often considered irrelevant to the work. One reason for this shift away from a consideration of the moral dimension in design is, as Levine, Miller and Taylor [2004] observe, that the concept of an ethics of architecture has become increasingly problematic over the last few centuries. This is because throughout the late nineteenth and early twentieth centuries philosophers persuasively argued that only human actions can have ethical connotations. Inanimate and non-sentient objects such as architecture are assumed to be without innate ethical capacity because they do not necessarily shape human responses, actions or behaviours. While in recent years the anthropocentric focus of ethics has been successfully challenged and broadened to include consideration of animals, plants and ecosystems, manufactured or synthetic objects have remained largely excluded from the ethical domain. This explains why works which are concerned with design and aesthetic qualities in buildings – such as Karsten Harries' *The Ethical Function of Architecture* [1997] – are in a minority, whereas texts on ethical decision making in architectural practice are more common [Wasserman, Sullivan, Palermo 2000; Spector 2001].

Despite this trend, in the last decade scholars have begun to raise new arguments about the merits of ethics in architectural design. One of the most interesting volumes published on this topic in recent years was Warwick Fox's edited compilation on *Ethics and the Built*

Environment [2000]. In this work a range of authors, under Fox's guidance, offer a series of proposals concerning the impact of buildings on social and physical environments. The book contains a strong call for architects to design in an ecologically sensitive way and to take account of the ultimate impact of their work. Authors in the work use both classical *virtue*-based arguments as well as evaluations of the impact of a building over time (what philosophers might call a *deontological* consideration) to support their propositions. Chapters in the book also offer important thoughts on the extent to which care should be taken in the aesthetic and social design of buildings. As editor of this volume, Fox is to be congratulated that the chapters remain typically balanced and avoid simplistic answers in favour of encouraging more detailed consideration of these topics in future research. For this reason Fox's return to some of this territory with his most recent book, *A Theory of General Ethics*, is noteworthy. Of even greater interest is that one of his aims in the new work involves reconnecting the synthetic or manufactured world to questions of ethics. As Fox observes,

> [...] if we think that it is possible to ask sensible ethical questions—sensible questions about the values we should live by—in respect of the human-constructed environment itself [...] then it follows that the central problem relating to the ethics of the human-constructed environment is that there presently isn't one! [Fox 2006: 48].

As the title implies, Fox's *A Theory of General Ethics* sets out to provide an overarching and consistent theory to connect and rationalise differences between the human world, the natural world and the synthetic world, the latter category including architecture. Initially in his new work Fox argues for the importance of a general theory of ethics in order to allow some sense to be made of multiple conflicting ethical views that are being propagated in the world today. It is the desire to offer a general theory that forces Fox to confront the classical division ethical philosophers have perpetuated between humans (and belatedly animals and the environment) and manufactured objects. This lack of ethical consideration of connection to

> ... the human-constructed environment represents the lack of an ethics in respect of what we might think of as the third main realm of our existence, that is, the realm of material culture (which includes all the "stuff" that humans make) as opposed to the biophysical realm (which includes eco-systems and the plants and animals that live in them) or the realm of symbolic culture (which is constituted by language-using human moral agents). ... [A]ny ethics that cannot directly address problems in this "third realm" is not even a candidate for a General Ethics [Fox 2006: 48].

At the core of Fox's reformulation and expansion of the classical ethical dilemmas to include the realm of material culture is his division of the world into three conceptual categories; fixed cohesion, responsive cohesion and discohesion. The first category, fixed cohesion, describes systems of any kind (social, natural, linguistic, structural, etc.) which are monotonous, constrained, externally ordered or moribund. The second category, responsive cohesion, describes systems that have loose boundaries, choice within limits, a degree of internal adaptability, and some limited predictability. The final category, discohesion, describes aleatory, arbitrary, anarchic, and unstructured systems. In a simplification of his position, Fox argues that in all cases the middle position, responsive cohesion, is at the core of a balanced existence, and that seeking such conditions in social, natural and manufactured environments is at the heart of a universal theory of ethics. By starting from this abstract, if thoughtful, position Fox is able to avoid many, but perhaps not all, of the criticisms that are typically levelled at Western ethical theory by non-Western cultures and religions. In essence, Fox is creating a metalanguage of ethical values that allows a wide range of issues to be logically and consistently illuminated. He clearly articulates the differences between inherent responsive cohesion and contextual responsive cohesion, and in doing so sets up a hierarchy that takes into account the impact of human, natural and manufactured objects on their local and wider context. For example, Fox states that,

> [a]lthough all forms of responsive cohesion are valuable, if conflicts arise between internal and contextual forms of responsive cohesion, as they often do, then considerations regarding the preservation and regeneration of contextual responsive cohesion are ultimately more important than considerations regarding the preservation and regeneration of internal responsive cohesion … [2006: 301].

Throughout his text, Fox uses architectural examples to explain a series of thought processes that could be used to determine if a building respects and serves the tenets of responsive cohesion and thus might be argued to be ethically sound. For this reason Fox's architectural arguments tend to value or privilege architecture which is in keeping with its context and responds to local conditions, materials, construction techniques and programs. On several occasions the "degree of contextual fit" of a building or object is suggested as the primary determinant of ethical value, all other things being equal. It is not surprising then that Fox references Christian Norberg Schulz's views on *genius loci* and Christopher Alexander's work on appropriate scale and ordered complexity, and regularly circles around the topic of Critical Regionalism. However, Fox avoids explicitly supporting the architecture that arises from such approaches because individual cases are far more complex, and he is aware that urban and rural settings call for different solutions to achieve responsive cohesion.

In conclusion, it is important to note that this book is not explicitly written for architects, but Fox's examples are clearly explained and most readers should have no difficulty following the majority of his propositions. The idea of a general ethical theory is a powerful one that at first appears beyond the scope of any book, let alone one readable by the intelligent layperson. Yet Fox elegantly and lucidly makes a case for just such a system and explains how it may be practically deployed. This is a stimulating and informative work that architects should read even if they must be wary of taking the theory of responsive cohesion too simplistically. This work does not reject the construction of singular works of architecture such as the Sydney Opera House, the Eiffel Tower or the Mole Antonelliana, but it does remind architects that contextual and environmental qualities should be at the forefront of our considerations.

Bibliography

FOX, Warwick. ed. 2000. *Ethics and the Built Environment.* London: Routledge.
HARRIES, Karsten. 1997. *The Ethical Function of Architecture.* Cambridge: MIT Press.
LEVINE, Michael P., K. MILLER, William TAYLOR. 2004. Ethics and Architecture. *The Philosophical Forum* **XXXV**, 2: 103-115.
SPECTOR, Tom. 2001. *The Ethical Architect.* New York: Princeton Architectural Press.

WASSERMAN, Barry, Patrick SULLIVAN, Gregory PALERMO. 2000. *Ethics and the Practice of Architecture.* New York: John Wiley

About the reviewer

Professor Michael J. Ostwald is Dean of Architecture at the University of Newcastle (Australia), a visiting Professor at RMIT University in Melbourne and a Professorial Research Fellow at Victoria University in Wellington (New Zealand). His recent books include *The Architecture of the New Baroque* (2006) and *Residue: Architecture as a Condition of Loss* (2007). In addition to his architectural qualifications, he holds the degrees of Doctor of Philosophy and Doctor of Science.

Book Review

Christy Anderson

Inigo Jones and the Classical Tradition

New York: Cambridge University Press, 2007

Reviewed by Sarah Clough Edwards

Baffy Cottege, Rosebery Road
West Runton, Norfolk NR279QW UK
sarah@talltrees84.freeserve.co.uk

Keywords: Inigo Jones, Andrea Palladio, English neo-Classicism, architectural education

As a key figure in English architectural history, Inigo Jones has already been the subject of considerable scholarship and publication. Christy Anderson's book, however, ably demonstrates the value of giving further consideration to well-studied and well-known figures. Whilst ostensibly a text about Inigo Jones, the underlying theme of the book is the evolution of an architect in seventeenth-century England, his persona, his training and his legacy. This is a fascinating notion, and one that reveals a wealth of information regarding the availability of mathematical and architectural texts and the pursuit of knowledge in the early seventeenth century.

The book is divided into seven chapters, each addressing a theme related to the overall discussion of Jones, his education and his architecture. This format results in a certain amount of overlapping in the text; the subject of measurement, for example, is raised in a number of chapters, as is the discussion of the link between Jones and the architect Andrea Palladio. The layout of the book is such, however, that whilst the text can be read from beginning to end in the traditional manner, the work does not adhere so strongly to a linear chronology that selective reading is impossible.

Chapter One, "Books and Buildings", is an introduction to the premise of the book and provides a guide to the many texts that formed the basis of an architect's knowledge. These works spanned the fields of science, mathematics, geometry, music and literature as well as architecture itself, and were written in a number of European languages. The introduction makes clear that the key to Jones's success was his scholarship, and that this scholarship went beyond that of his contemporaries.

The following three chapters concentrate on the library of architectural, mathematical and scientific texts accumulated by Jones, and examine the way he employed and understood these works. In Chapter Two, "The Famous Mr Jones", Anderson considers Jones's pursuit of fame and the importance of key patrons in his early career, whilst also examining his early architectural commissions and designs. Anderson demonstrates the importance of Jones in developing the notion of the gentleman architect, underlining that Jones worked hard to create and promote a particular image. For this reason, however, the experience and education of Jones set him apart from English architects in general. Chapter Three, "Building a Library", underlines the importance of accumulating books in the seventeenth century and the difficulties faced in collecting such texts. In addition to a book collection, architects also required various, and highly expensive, tools employed for the survey and measurement of sites and structures. For this reason a wealthy patron was another vital element for an architect's professional development. Chapter Four, "Conversations with the Dead", examines the way in which Jones used the text of Palladio and other architectural writers in conjunction with first-hand study of architecture. This discussion considers the importance of measurement to the architect in his study of the architecture of others, and the way in which such information was recorded.

Chapter Five, "The Hand of Inigo Jones", comprises an examination of handwriting technique and the transmission of information and ideas through the writing and drawing of Jones. It also considers the dialogue between the living Jones and the dead Palladio through Jones's annotation of Palladio's text. Chapter Six, "A More Masculine Order", examines Jones's quest to find an architectural style which embodied the best of his predecessors' and mentors' work, gleaned from their texts and architectural output, but which also demonstrated a new, idiosyncratic, and English style. This pursuit underlines the fact that whilst Jones drew extensively on the work of Palladio, at no time was his work simply a regurgitation of Palladian models. Palladian principles can be easily discerned in the work of Jones, particularly in the modular design of the Banqueting House, yet other designs by Jones, such as that for the termination to the tower of Old St Paul's Cathedral, may contain Palladian elements such as *serliana* but could never be mistaken for a Palladian work. Chapter Seven, "Practices", charts the relationship between books and buildings in the work of Jones, examining the process by which an architect's theoretical studies are translated into working practice. This chapter contains detailed studies of three Jones projects, The Banqueting House and Whitehall Palace, Old St Paul's Cathedral, and Covent Garden.

The conclusion of the text considers the legacy of Jones and his importance as a figurehead in the development of the English neo-Classical and Baroque. Anderson's discussion underlines the pivotal importance of Jones in the development of the English neo-Classical style. Jones's legacy, in part, lay in the fact that having travelled to Italy and spoken to those who knew Palladio personally, he could be regarded as a direct link between the Italian architect and the English neo-Classical style. In this sense, Anderson argues, the figure of Jones provided a line of direct descent between England and Italy.

One of the achievements of this book is its demonstration that to consider the architecture of Inigo Jones solely in relation to that of Palladio is to do a great disservice to the English architect. Jones studied extensively in Italy, examining the work of many architects in great detail and giving due consideration to the writings of Alberti, Serlio and others. The fraction of Jones's library which survives is evidence of the breadth of his scholarship and the many influences on his artistic development. That said, Palladio remains the central figure of influence in Anderson's text, and certainly less discussion is given to other influential architects, such as Michele Sanmicheli and Giulio Romano.

Whilst mathematical measurement is central to the theme of the text, and was central to the architectural prowess of Jones, the process of learning and undertaking mathematical measurement is not examined in the same detail as is the acquisition of knowledge through reading and on-site examination. Whilst it is noted that Jones employed the geometrical square, for example, no further information regarding this device is provided. This is regrettable, since the depth of information regarding literary sources for Jones would have been enhanced by a parallel description of the tools of the trade. Jones kept lists of building measurements he had taken, in his copy of Palladio's *Quattro Libri*. Anderson provides a section of these measurements, and it is interesting to note the paradox of Jones carefully recording distances measured in the less than accurate medium of paces. This difficulty in obtaining accurate measurements was compounded by the disparity between English measurements and the various Italian systems of measurement such as Roman palms and Venetian feet. This complexity of measurement is a subject not fully addressed in this text.

Overall, Anderson makes very clear the massive importance of the written word in the pursuit of knowledge, and architectural skill, in the seventeenth century. For this reason it is surprising that Anderson does not question the fact that Jones himself did not seek to immortalise his work in an architectural treatise similar to the *Quattro Libri* of Palladio. This book acknowledges that among his predecessors and contemporaries Ingio Jones stood apart, in that no other 'so closely linked their activities as designers with reading and the study from books'. It is evident that scholarship was of fundamental importance to Jones and yet, despite a personality described by a contemporary as 'conceited and boastful', he did not formalise his theories for future generations. Perhaps the best answer to this question is that Jones saw his constructed architecture as his legacy. It is fascinating to learn that Jones's words were held in such regard that already by 1715 Giacomo Leoni sought to publish a transcript of Jones's annotations to the *Quattro Libri* in conjunction with the *Quattro Libri* itself. This project was never properly brought to fruition, but the illustrations within Anderson's text of some of these annotated sheets are fascinating and a clear demonstration of the value of such an undertaking.

Although this text does not examine the architecture of Jones, ultimately its primary purpose is to provide an analysis of what Jones thought about architecture, how he developed designs, and an examination of the origins of his ideas and theories. The central theme, which addresses the library of Jones and the pursuit of knowledge through the available literature of the era, is compelling and provides insight into the commitment and scholarship of Jones himself. In this work Anderson has constructed a fascinating historical text that provides much insight into the world in which Inigo Jones lived and worked.

About the reviewer

Sarah Clough Edwards received her Ph.D. in architecture history from the University of Reading. She is now associated with the University of East Anglia, where she lectures on architectural history. Her Ph.D. research concerned the design of the convent of S. Chiara in Urbino and its architect, Francesco di Giorgio Martini of Siena. Several articles drawn from this research are now in the process of being published. Her current research is concerned with the architectural evolution of the staircase in the Renaissance period and its symbolic utilisation within the complex social mechanisms of the Italian courts. Edwards has also presented a number of conference papers on her research.

Book Review

Corinna Rossi

Architecture and Mathematics in Ancient Egypt

Cambridge: Cambridge University Press, 2004

Reviewed by Sylvie Duvernoy

Keywords: Egyptian architecture, Egyptian mathematics

Via Benozzo Gozzoli, 26
50124 Florence ITALY
sduvernoy@kimwilliamsbooks.com

A book bearing such a title surely will arouse the curiosity of the readers of the *Nexus Network Journal*, especially those engaged in the study of ancient architecture. So far, the scientific study of the reciprocal interactions between mathematics and architecture has produced a vast amount of articles and papers, but there are not yet many books in which every chapter is dedicated only and entirely to this topic, especially as far as pre-classic antiquity is concerned.

Architect and Egyptologist Corinna Rossi presents the results of the research that she carried out while preparing her Ph.D. at Cambridge University under the guidance of Prof. Barry J. Kemp. Her main purpose is to understand how Egyptian architects used mathematical concepts in the process of designing and building architectural monuments. The research methodology that she defined for the achievement of her project consists in approaching ancient Egyptian architecture from two different standpoints, and in combining the knowledge of architectural historians, who mainly focus on building remains, with the knowledge of Egyptologists, who mostly study textual and figurative (and other) remains. The book is therefore divided in three parts. The first part is an overview of the historical theories suggested (until very recently) to explain the proportions of ancient Egyptian architecture. Part two is dedicated to the analysis of all the surviving archaeological evidence of the planning and building processes in ancient Egypt. Part three is the attempt to reconcile the architectural and archaeological approaches to the study of the relationship between architecture and mathematics.

In her attempt at highlighting interactions between Egyptian science and architecture, Corinna Rossi's most valuable effort lies in the definition of a strict methodology – as objective as possible – based on comparative analysis, capable of producing reliable scientific conclusions. The first "mathematical" interpretations of Egyptian architecture began in France in the nineteenth century, and were prompted by the illustrations published in the *Description de l'Egypte* authored by the scientific escort of Napoleon during his oriental campaign. In part one of her book, Corinna Rossi reviews various interpretations – among which the ones by Viollet-le-Duc (1863), Choisy (1899) and Badawy (1965) – showing how they all tend to subjective conclusions, as they are based on some anachronistic concepts. In the preface Rossi writes:

> These theories do not necessarily provide any useful information about the ancient culture to which they are supposed to refer, but on the other hand they may play an

1590-5896/08/010203-4 DOI 10.1007/ S00004-007-0064-8
© 2008 Kim Williams Books, Turin

important role in the study of the culture and the historical period that produced them – that is, Europe in the last two centuries.

In fact, the best research methodology for analysing historical design principles from a geometrical/arithmetical standpoint consists in the comparison between architecture and the contemporary mathematical knowledge. The manipulation of proportional systems and geometric patterns has to be rigorously coherent with the historical scientific context. Although this concept might seem obvious, it is paradoxically rather recent. "Thinking of it, this constraint should have imposed itself long ago" says Pierre Gros, French scholar of Roman architecture [1995, 21]. But it did not. This attitude started to spread among scholars only in the last few decades, in parallel to a greater accuracy in survey operations and rigour in digital representation.

In speaking of Egyptian mathematics, many notions have to be redefined. Corinna Rossi's outline of the mathematical knowledge in ancient Egypt is drawn from the many comments written by historians of mathematics on the four main extant original documents: the "Rhind" and "Moscow" mathematical papyri, the Kahun Papyri, and the "Egyptian Mathematical Leather Roll". Following her bibliographic sources, Rossi asserts that the "Egyptian triangle" – right-angled, with sides of 3 and 4 and hypotenuse of 5 – was given its name by the Greeks, and that no written Egyptian document really proves that it had a special function or meaning in early Egyptian culture. Viollet-le-Duc, in his *Entretiens sur l'Architecture* [1863], calls "Egyptian" the isosceles triangle with base 8 and height 5. According to him, this particular triangle comes from the diagonal vertical section of a pyramid having a height of 5 and whose vertical cross-section is an equilateral triangle. Those – very – approximate ratios suggested to some nineteenth- and twentieth-century scholars that the Egyptians were already aware of the Golden Section and the Fibonacci arithmetical series, a hypothesis that Rossi firmly rebuts.

Every architect engaged in the search for hidden geometrical patterns in historical architecture sooner or later bumps into the realization that some of the geometrical figures that appear while analyzing the survey drawings may be mere consequences of other primary layouts. Reconstructing the design process adopted by the original designer consists in recognizing the difference between intention and coincidence, far from any cultural influence tending to lead towards some specific "preferred" geometrical patterns.

The second part of the book focuses on Egyptian architectural documents. Corinna Rossi does not intend to draw a picture of the history of Egyptian architecture but she analyses the archaeological remains of documents related to design and construction. She lists twenty architectural sketches and drawings, eight full-size geometrical sketches, and five architectural models that have survived and are kept in various museums. The overall quantity of these documents may seem rather abundant compared with what comes from the Greek and Roman cultures; however, the oldest of these remains are dated from the third dynasty (2686-2600 B.C.) and the more recent are from the Roman Period (30 B.C.-395 A.D.), which means that they spread over a period of time which covers nearly 3000 years. Therefore, what appears at first sight to be a fair amount of documentation is, in fact, a very limited quantity of material compared to the length of duration of the civilization about which it is supposed to give us information. And therein lies the first difficulty of interpretation. Since it lasted for so many centuries, Egyptian civilization cannot be considered as a whole, unified cultural period; how, therefore, are scholars to take into consideration evolution and progresses in scientific knowledge while analyzing remains from different ages? The most productive age of extant architectural documents is the period known as the "New Kingdom", corresponding to the eighteenth, nineteenth, and twentieth dynasties (1550-1069 B.C.). Thirteen out of the twenty architectural sketches and drawings listed by the author come from that period. In fact, the most interesting chapters of the book are the ones related to the study of the Royal Tombs, in the Valley of the Kings, in Thebe (nineteenth and twentieth dynasties). An additional nineteen documents on construction processes, all related to these tombs, makes possible direct comparison between the written (or drawn) evidence and the actual monumental remains. Bearing in mind that the Rhind Papyrus is dated from sometime around 1650 B.C., we have here a situation where mathematical and architectural documents, and buildings are all roughly contemporary, a particularity which the author does not really underline, and from which no specific conclusion is drawn.

Corinna Rossi chose to include in her overview of the Egyptian architectural documents the "building texts" of the two Ptolemaic temples of Edfu and Dendera, the first of which was built 237-142 B.C., the second 54-20 B.C. She says,

> Even if the architecture built in Egypt in the last three centuries B.C. and then under the Romans is strictly related to the ancient tradition of the country, it cannot be excluded that foreign influences combined with the old traditions [Rossi 2004: 173].

In fact, at the time when the construction of Edfu began, the "Golden Age" of Greek geometry was reaching its acme, and by the time the Dendera temple was completed, Archimedes had been dead for two centuries and Vitruvius was writing his treatise on architecture. Therefore, relating these two temples mostly to the ancestral Egyptian mathematical knowledge and to the ancient tradition of the country seems quite inappropriate. And surely it would be interesting to envision the process of influence between Egypt and Western culture in the opposite direction. Egypt gave and transmitted much knowledge to occidental countries before receiving back foreign influences. Since many of the early great Greek geometers studied in Alexandria, from Thales (600 B.C.) to Euclid (300 B.C.), the study of Egyptian Ptolemaic architecture could probably provide many clues about progress in Egyptian mathematical knowledge and about early Greek geometry, and thus about the transition from Egyptian to Greek relationships between mathematics and architecture.

In the final part of the book, the author applies her method of analysis to the study of the most fascinating architectural typology of ancient Egypt: pyramids. So much has been said about pyramids that it seems impossible to say more, but in fact, after discarding all the superficial, fantastic and esoteric writings, very few real scientific works are left. Roger Herz-Fischler [2000], for instance, has dedicated a full book to all the theories suggested so far by the Great Pyramid of Khufu alone. Rossi's conclusion is that "they are representative of the modern culture which generated them, rather than of the ancient culture to which they refer" [2004: 201].

Corinna Rossi's personal contribution to the study of pyramids is to try to unveil some typological geometric rules, taking into account an historical evolution of the shape throughout the centuries. It is indeed a risky proposal. However, the result is rather convincing, especially because the conclusions are discussed objectively, pointing out the limits of such an inquiry, and also because the author does not try to prove more than what her documentation allows her to do. The author lists about a hundred monuments coming from the very first so-called "bent pyramid" built by Snefru, up to the end of the thirteenth dynasty. However, only sixty-eight of the pyramids listed are completed with measures regarding both the side of the base and the angle of the slope. Corinna Rossi did not measure all the pyramids herself, but collected numbers from various data-bases. The origin of the data is mentioned for every monument, together with the degree of reliability of each survey.

From her collection of measurements, Rossi establishes fourteen different numerical ratios between the base and the height of the pyramid, the ratio which determines the "*seked*" of the pyramid, that is, the slope of its faces. Some of these numerical ratios are more recurrent than others. They are the ones that define the most regular geometrical figures: the equilateral triangle of the cross vertical section, or of the pyramid face itself. But other patterns are discussed that highlight Egyptian skills in combining geometry and arithmetic. Every scholar in Egyptian architecture can find here an interesting and innovative presentation of already available information that had never been assembled in such a way previously. Some famous Egyptologists produced extremely precise information on single monuments, or on single periods, but the value of Corinna Rossi's work is to draw a full overview of the pyramid typology. Comparative analysis is a powerful tool of inquiry. Mark Wilson Jones has written:

> Exceptions and compromises are only to be expected of any grouping of the products of human creativity, and if half to two-thirds can be seen to conform to a pattern, that is quite sufficient to demonstrate that a certain procedure existed and was reasonably if not universally popular [Wilson Jones 2006].

Not one single photograph, even in black and white, illustrates Corinna Rossi's book: only drawings, schemes and sketches. No concession was made to seduce a non-academic readership! However, the architect reader, professionally trained to examine and criticize his peers' drawings, cannot help but notice that many illustrations are labelled "drawn from XXX", indicating that they are not first-hand but second-hand drawings. Copies from copies never produce scrupulous representations. Some photographs or reproductions (even at the cost of paying some copyright fees!) would have helped to understand the level of interpretation of the original transcriptions, and would have granted the book the touch of *pathos* that it lacks.

References

BADAWY, Alexander. 1965. Ancient Egyptian Architectural Design. In *Near Eastern Studies 4*. Berkeley, University of California Press

GROS, Pierre. 1995. Les illustrations du *De Architectura*. In : *Les littératures techniques dans l'antiquité romaine* C. Nicolet, ed. Genève : Vandoeuvres.

HERZ-FISCHLER, Roger. 2000. *The shape of the Great Pyramid*. Waterloo, Ontario: Wilfrid Laurier University Press. (Reviewed by Mark Reynolds in the *Nexus Network Journal* 3, 2: 197-200.

VIOLLET-LE-DUC. 1863. *Entretiens sur l'Architecture*. Paris

WILSON JONES, Mark. 2006. Ancient Architecture and Mathematics: methodology and the Doric temple. Pp. 149-170 in *Nexus VI: Architecture and Mathematics*, Sylvie Duvernoy and Orietta Pedemonte, eds. Turin: Kim Williams Books.

About The Reviewer

Sylvie Duvernoy is Book Review editor of the *Nexus Network Journal*.